Books in the Alpha Wolves of Yellowstone Series by Rick McIntyre

The Rise of Wolf 8:
Witnessing the Triumph of
Yellowstone's Underdog

The Reign of Wolf 21:
The Saga of Yellowstone's
Legendary Druid Pack

The Redemption of Wolf 302:
From Renegade to
Yellowstone Alpha Male

The Alpha Female Wolf:
The Fierce Legacy
of Yellowstone's 06

RICK MCINTYRE

Foreword by **MARC BEKOFF**

THE Reign of Wolf 21

THE SAGA OF YELLOWSTONE'S LEGENDARY DRUID PACK

GREYSTONE BOOKS
Vancouver/Berkeley/London

First paperback edition 2022

23 24 25 26 27 6 5 4 3 2

Greystone Books Ltd.
greystonebooks.com

Cataloguing data available from Library and Archives Canada
ISBN 978-1-77164-996-4 (pbk.)
ISBN 978-1-77164-524-9 (cloth)
ISBN 978-1-77164-525-6 (epub)

Editing by Jane Billinghurst
Copyediting by Rowena Rae
Proofreading by Meg Yamamoto
Maps by Kira Cassidy
Cover and text design by Fiona Siu
Cover photograph of a black wolf by Michelle Lalancette
Printed and bound in Canada on FSC® certified paper at Friesens. The FSC® label means that materials used for the product have been responsibly sourced.

Greystone Books thanks the Canada Council for the Arts, the British Columbia Arts Council, the Province of British Columbia through the Book Publishing Tax Credit, and the Government of Canada for supporting our publishing activities.

Canada

Greystone Books gratefully acknowledges the xʷməθkʷəy̓əm (Musqueam), Sḵwx̱wú7mesh (Squamish), and səlilwətaɬ (Tsleil-Waututh) peoples on whose land our Vancouver head office is located.

CONTENTS

Map of Northeast Yellowstone National Park *viii*
Foreword by Marc Bekoff *x*
Prologue *xiii*
Previously in Lamar Valley *xiv*

PART I: 2000
Range Map *2*
Wolf Charts *3*
1 First Winter *5*
2 Insurrection *15*
3 Counting Pups *31*
4 The Pups Grow Up *43*

PART II: 2001
Range Map *52*
Wolf Charts *53*
5 It's Complicated *54*
6 Separate Dens *62*
7 The Biggest Pack *70*
8 The Battle of Lamar Valley *81*

PART III: 2002
Range Map *92*

Wolf Charts *93*
9 The New Packs *96*
10 The Battle of Hellroaring Creek *105*
11 Dens and Pups *112*
12 Wolves and Ravens *124*
13 Invasion and a Separate Peace *136*
14 253's Incredible Journey *143*

PART IV: 2003
Range Map *150*
Wolf Charts *151*
15 Enter Wolf 302 *154*
16 Raising Pups *167*
17 Coexistence *180*

PART V: 2004
Range Map *196*
Wolf Charts *197*
18 January *200*
19 February *205*
20 March and April *213*
21 May *220*
22 June *224*
23 The Quest *230*

Epilogue *236*
Acknowledgments *237*
Author's Note *239*
References *240*
Index *243*

"I was thinking about my June and how we share everything. When one of us is cut, we both bleed. When one of us is sick, we both hurt. When you are married you are one. You are bonded to someone that you love and pull together as one."

JOHNNY CASH SPEAKING ABOUT HIS WIFE IN THE DOCUMENTARY *JOHNNY CASH AT FOLSOM PRISON* (2008)

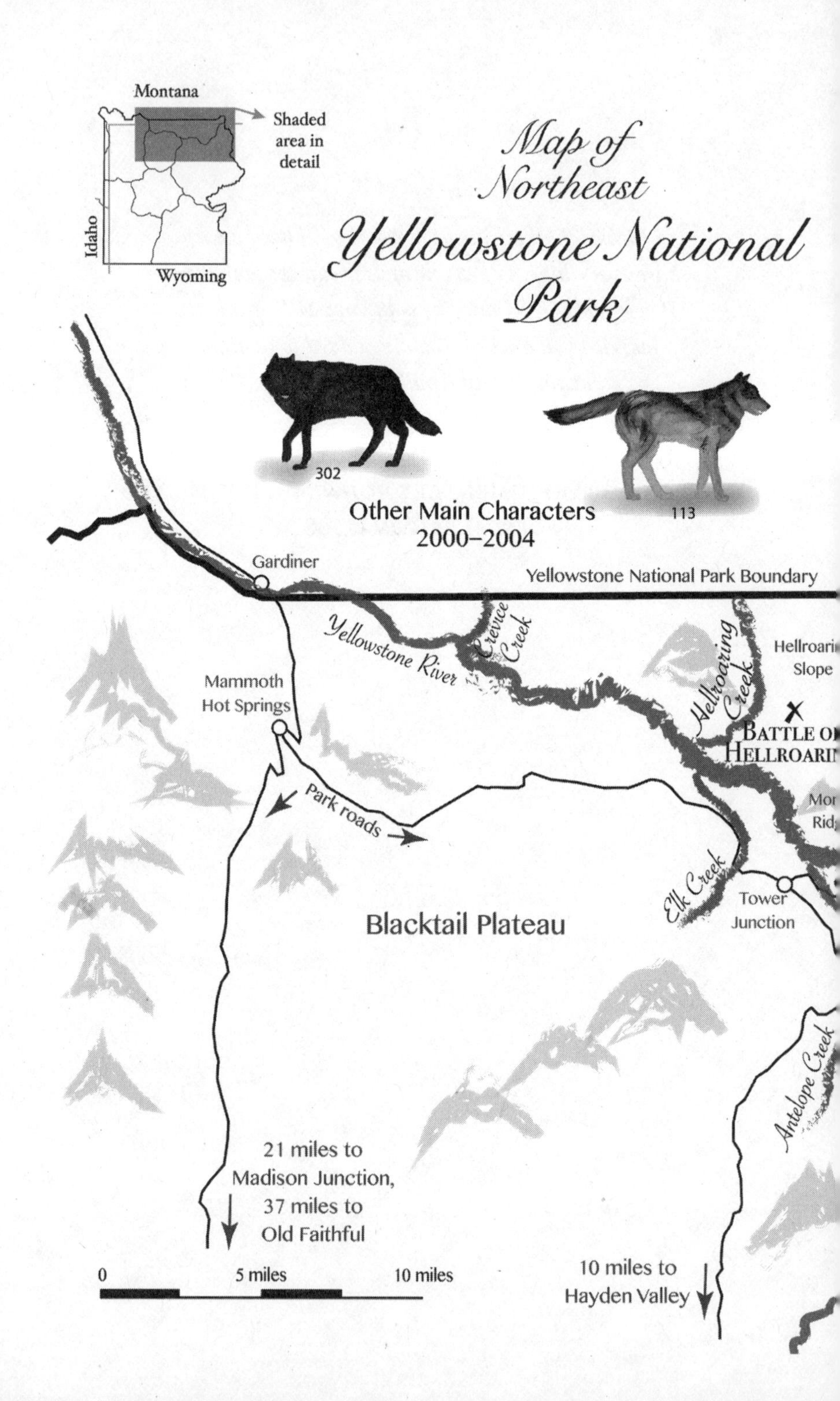

Montana
Shaded area in detail
Idaho
Wyoming
Map of Northeast Yellowstone National Park
302
113
Other Main Characters 2000–2004
Gardiner
Yellowstone National Park Boundary
Yellowstone River
Crevice Creek
Hellroaring Creek
Mammoth Hot Springs
Park roads
Elk Creek
Tower Junction
Blacktail Plateau
Antelope Creek
21 miles to Madison Junction, 37 miles to Old Faithful
0
5 miles
10 miles
10 miles to Hayden Valley

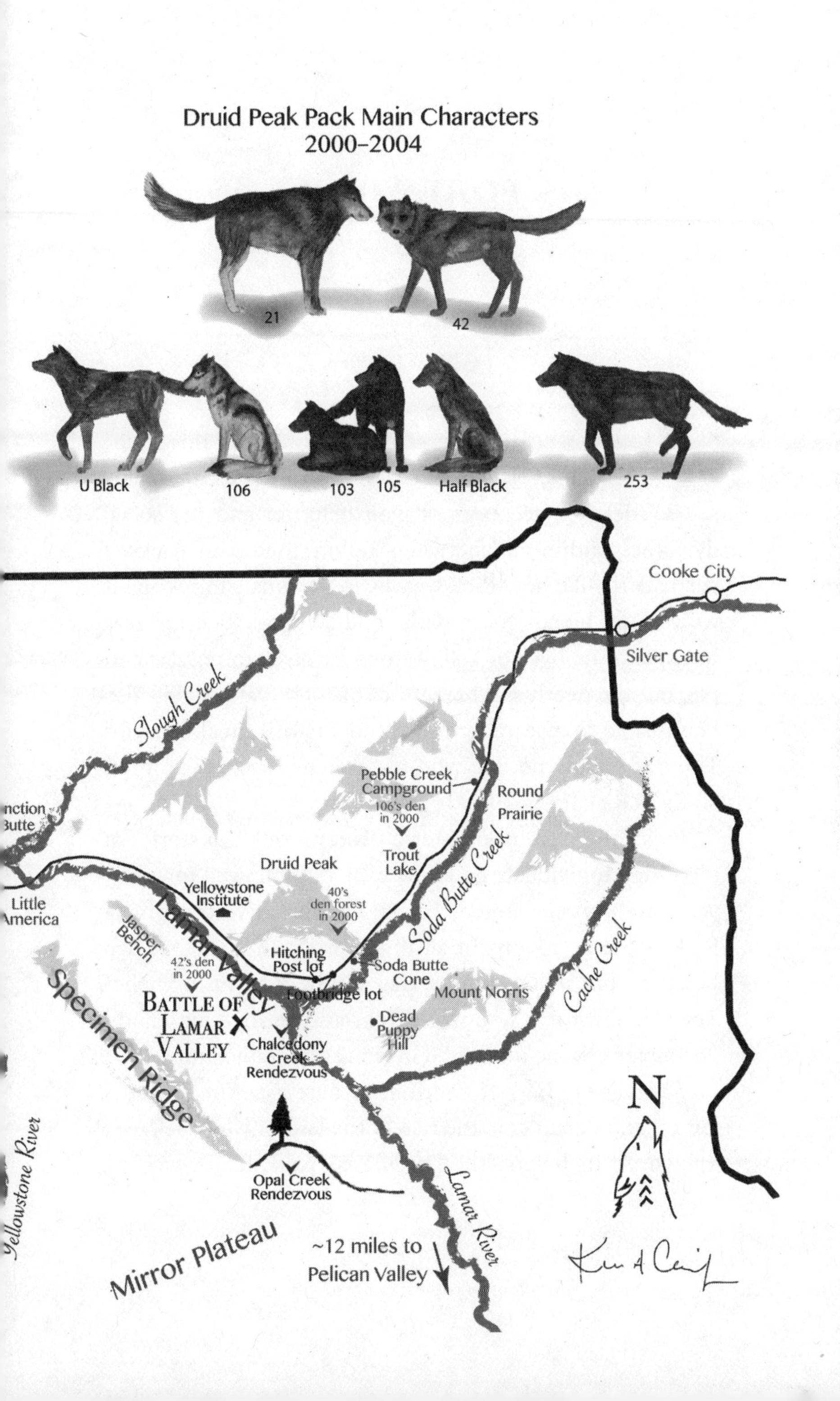

Druid Peak Pack Main Characters
2000–2004
21
42
U Black
106
103
105
Half Black
253
Cooke City
Silver Gate
Slough Creek
Pebble Creek
Campground
Round
Prairie
106's den
in 2000
Trout
Lake
Druid Peak
Soda Butte Creek
Yellowstone
Institute
Little
America
Lamar Valley
40's
den forest
in 2000
Jasper
Bench
Hitching
Post lot
Soda Butte
Cone
Cache Creek
42's den
in 2000
Footbridge lot
Mount Norris
Specimen Ridge
Battle of
Lamar
Valley
Dead
Puppy
Hill
Chalcedony
Creek
Rendezvous
N
Yellowstone River
Opal Creek
Rendezvous
Lamar River
Mirror Plateau
~12 miles to
Pelican Valley

FOREWORD

AS I LEARNED when I read the first book in this series, *The Rise of Wolf 8,* Rick McIntyre writes incredibly detailed accounts of wolf behavior and the social dynamics within and between Yellowstone wolf packs. I particularly like how Rick weaves in personal stories about wolves and blends these with solid science. Rick's goal is to explain the lives of Yellowstone's wolves to regular people, but the depth and breadth of his observations will also be valuable to researchers and students of animal behavior. There is simply no one who has watched wolves as intensively as Rick has.

Rick's meticulous long-term observations and stories of identified individuals he knew well and followed for many years are unprecedented. For many years, I've thought of Rick as the go-to guy for all things wolf, and the first two books in his Yellowstone series amply confirm my belief. They are must-reads, to which I'll return many times, and I encourage anyone interested in wolves to do the same.

The Reign of Wolf 21 continues where *The Rise of Wolf 8* left off and documents the rise of the largest wolf pack ever known. At its height, the Druid Peak pack, led by wolf 21, comprised thirty-eight wolves and held sway over an enormous territory in Lamar Valley. The intrepid alpha male achieved all this by being fearless in battle, never backing down, never killing a rival wolf, and, even more importantly, having an equally loyal, fearless, and wise companion by his side, wolf 42. The story of their devotion to each other is at the heart of this book.

Wolf 21 was raised by his adoptive father, wolf 8, the subject of the first book in the series. The two had an especially close relationship, and later in his life, 21 exhibited many of the leadership skills passed down to him by his mentor. From watching 8 and then his adopted son, 21, Rick learned how multiple adult wolves in a family cooperate to raise and feed their young and protect them from threats such as grizzlies and rival wolf packs.

A particular interest of mine in both wild and domesticated canids (that is to say, members of the dog family) is their capacity for play. Rick makes it clear in this book that 21 was one of the most playful wolves he has ever had the privilege to watch. The big alpha male (21 had inherited his size from his biological father, wolf 10, who was an impressively large and strong wolf) loved to engage in games with his pups and would let little pups beat him in wrestling matches. Rick got the sense that 21 liked to pretend he was a low-ranking wolf when he was with younger pack members, a type of role reversal. 21's concept of being an alpha male was the exact opposite of what we think of in humans as an aggressive, dominating alpha male personality.

Rick also studied the many types of games pups played among themselves and saw how those games, such as chasing and wrestling, prepared them for their adult responsibilities of hunting and protecting their families from other wolves. All of Rick's observations sound just like what dogs do when they're allowed to run freely and play with one another or alone. Years ago my students and I observed similarities between the play of wild coyotes and that of domestic dogs. Rick's reports extend these similarities to wolves, as well.

Rick's descriptions of the wolves of Yellowstone remind me of Dr. Jane Goodall's early groundbreaking research on wild chimpanzees in which she named each individual and wrote about their unique personalities, a practice for which she was initially criticized by her professors, many of whom had never seen a wild animal of any type. Of course, her critics were totally incorrect as has been shown by subsequent research on chimpanzees and a wide variety of other animals.

As Rick once heard someone say, "It is hard to hate someone if you know their story." I think that after you have read the tales of drama, courage, and devotion in this book, you will agree.

MARC BEKOFF
Boulder, Colorado

PROLOGUE

21 WAS MISSING. HE normally was with his family every day, so his disappearance was troubling. 21 was an old wolf. He had lived nine years, about twice as long as an average wolf in Yellowstone. A month went by without any sightings of him.

Then an outfitter found a dead wolf and a radio collar in the mountains above Lamar Valley. The man turned the collar in to a ranger who passed it on to me. The collar was 21's.

A group of Wolf Project staff rode the steep trail up Specimen Ridge. We found 21 curled up on a low hill in a high-elevation meadow. The site where he lay overlooked Lamar Valley, where he had been the alpha male of the Druid Peak pack for over six and a half years.

For a long time, I tried to figure out why 21 had left his pack and used the last of his prodigious strength to travel to that meadow. He must have had a reason to go up there alone.

Years later inspiration came to me, a motive for his behavior. I first explained what I had come up with to a friend and when I finished, she began to sob. I asked what was wrong and after taking a few moments to compose herself, she said, "Why can't I find a man like 21?" Let me tell you why she said that.

PREVIOUSLY IN LAMAR VALLEY

IN THE FIRST book of this series, *The Rise of Wolf 8: Witnessing the Triumph of Yellowstone's Underdog*, I told the stories of the wolves brought down from Canada as part of Yellowstone National Park's 1995 wolf reintroduction program.

Among the early arrivals, the Crystal Creek pack had two adults and four male pups. The smallest pup was wolf 8, who was bullied by his brothers while the family was being held in their acclimation pen. After the pack was released into the wild, 8 unexpectedly proved himself when he stood up to an angry grizzly and helped save his siblings.

Another family, the Rose Creek pack, denned shortly after they were released from the pen where they had been held when they first arrived in the park. The alpha female, wolf 9, gave birth to eight pups. Their father was illegally shot and killed, making 9 a single mother with little chance of keeping her pups alive. A team of wolf biologists captured her and the pups and put them back in an acclimation pen, intending to release them in six months.

On the day of their release, wolf 8, who was then eighteen months old, about sixteen in human years, happened to come into that area. He spotted some of the Rose Creek pups and befriended them. That led to the pups' mother accepting 8 into her pack as the new alpha male, a promotion that would involve heavy responsibilities for the young wolf.

That winter, as 8 was helping to raise the pups he had adopted, four new packs were brought to Yellowstone. The Druid Peak pack's alpha male, wolf 38, was so strong that he tore apart his metal transport kennel. After they were released, the Druids attacked 8's original family, killed their alpha male, and took their territory away from them. The surviving Crystal Creek wolves fled south and found a new valley to settle. The group was later renamed Mollie's pack to honor Mollie Beattie, director of the U.S. Fish and Wildlife Service, who died the year after the reintroduction began. Mollie had been a strong supporter of wolf recovery and had helped carry some of the original Crystal Creek wolves to their pen in 1995.

In the spring of 1996, wolves 8 and 9 had pups together. One day 8 saw a pack of wolves charging downhill at his family. It was the Druids, led by their alpha male, the huge wolf who had killed 8's father. Without hesitation, 8 ran at the much bigger male, determined to protect the young wolves he was raising. The two alpha males had an all-out fight and underdog 8 defeated 38. He beat him up, then chose to spare his life and let him run off.

Those young wolves, including wolf 21, witnessed 8's courageous victory. 21 was raised and mentored by 8 for two years. He learned through example how to hunt, how to raise pups, and how to fulfill an alpha male's responsibilities to his family.

21 left home in the fall of 1997 when he was two and a half years old, about twenty-four in human years. He joined the rival Druid pack as their alpha male after the death of 38 and raised 38's pups like 8 had raised him and his siblings.

When 21 joined the pack, the Druids were led by a violent and domineering alpha female. Wolf 40 eventually drove her mother and one of her two sisters out of the pack. Her remaining sister, 42, put up with 40's bullying and abuse for years. We suspected she killed 42's pups two years in a row, in 1998 and 1999, to ensure the pack's resources were directed to her own litter.

From the first day they met, 21 and 42 seemed drawn to each other, possibly because their personalities were so similar. As I watched the two wolves over the next few years, I couldn't help thinking how much better their lives would be if something were to happen to 40. In the spring of 2000, I witnessed a dramatic series of events that radically restructured the Druid pack.

PART I

2000

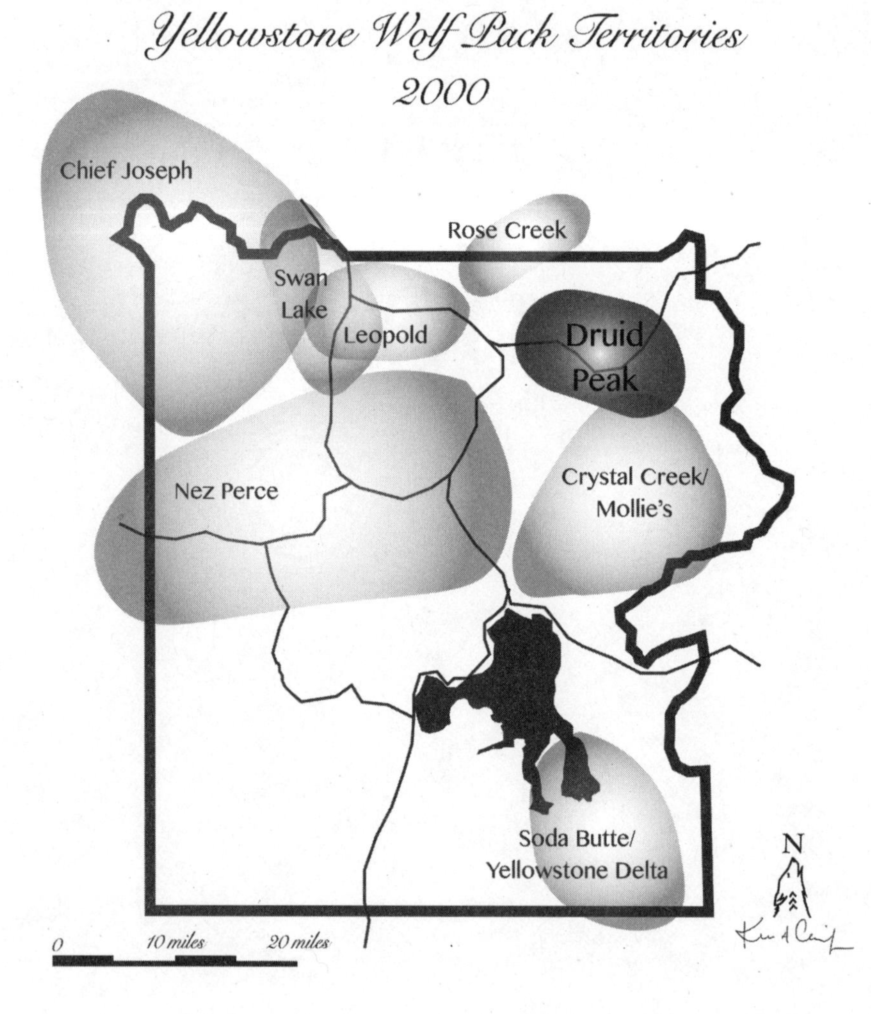
Yellowstone Wolf Pack Territories
2000
Chief Joseph
Rose Creek
Swan Lake
Leopold
Druid Peak
Nez Perce
Crystal Creek/ Mollie's
Soda Butte/ Yellowstone Delta
N
0
10 miles
20 miles

PACKS INCREASE AND decrease in size over the course of a calendar year. These charts show the main pack members in any given year. M=male and F=female. An asterisk (*) indicates a female thought to have denned. The pack of origin for wolves joining from other packs is indicated in parentheses the first time a wolf is introduced. Squares indicate adults and yearlings. Circles indicate pups.

Druid Peak Pack 1999

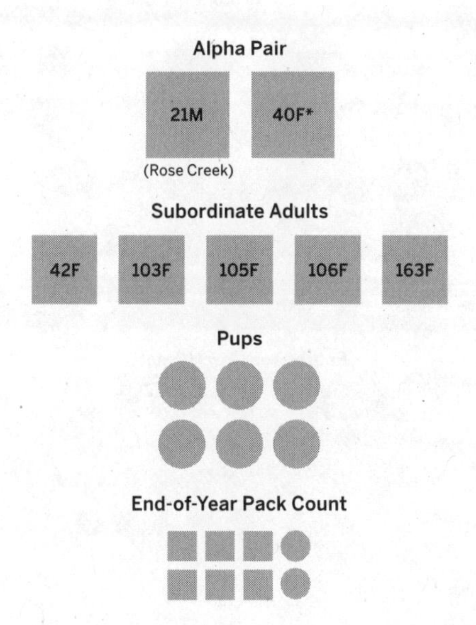

Druid Peak Pack 2000

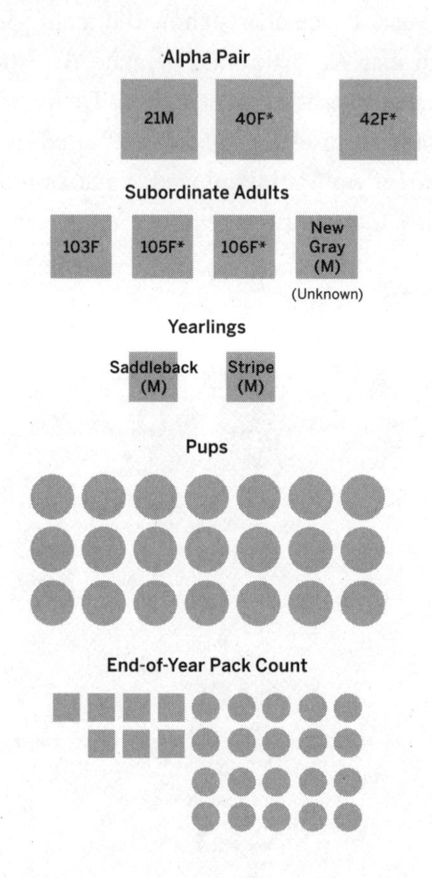

1

First Winter

FROM 1994 THROUGH 1997, I had worked only summers in Yellowstone, then left every fall for a park naturalist job in Big Bend National Park in west Texas. In the spring of 1998, I transferred from my naturalist position in Yellowstone to a summer job with the Wolf Project and helped to research and monitor the wolf population in the park. I also gave wolf talks for the public and helped visitors see wolves and understand their behavior. Starting in the spring of 1999, I worked for the Wolf Project year-round. Since my job involved being outside every day, I got concerned as the winter of 1999/2000 approached. I had grown up in New England so had experienced harsh winters, but having worked in desert parks for the last twenty-two winters, I was worried I had lost any previous ability to endure extremely frigid temperatures. In January 2000, I got to experience what winter weather is like in Yellowstone. The official temperature got as low as minus 33 Fahrenheit (–36

Celsius) and thick snow covered the terrain. One day several of us had to trudge one and a half miles uphill through deep snow to examine a bull elk the wolves had killed. It took us four exhausting hours.

My concerns about losing my resistance to cold weather turned out to be justified. I had a hard time keeping warm while doing my job, which often involved standing in one spot and watching wolves for hours at a time. Luckily I discovered hand warmers, chemical packs that heated up when shaken. I also used toe warmers that had an adhesive side so I could stick them to the bottom of my socks. I could not have survived the cold winters in Yellowstone without those aids.

The hand and toe warmers helped, but on really cold days my core body temperature was a problem no matter how many layers I put on. As I got cold and started to shiver, the wolves often were curled up and sleeping on the snow, perfectly adapted to the subzero conditions, thanks to the insulating qualities of their thick winter coats. If I couldn't warm up by walking around, I had to go back to my car, turn up the heater, and get back to normal before heading out again.

The Druid pack numbered eight that January: the alpha male, 21; the alpha female, 40; her sister, 42; three young adult females, 103, 105, and 106; and two gray pups. One of the pups had a saddlelike patch of dark fur over his shoulders, while the other one had prominent black stripes along his back. We referred to them as Saddleback and Stripe. Now eight months old, the pups had the stamina to keep up with the adults as the family ranged far and wide throughout their large territory in Lamar Valley in the northeastern section of Yellowstone.

As I watched the Druids that January, I learned how fussy wolves are in making beds in the snow, especially 42. One cold morning, I watched her spend two minutes fastidiously pawing out a depression in the deep snow, then tramping down the bed by making several tight circles before finally lying down and curling up. The crater kept her out of the wind and minimized heat loss. As she made those circles in the snow, I thought of how dogs, even poodles inside warm luxury Manhattan apartments, go through similar motions when they get ready to bed down, a remnant of their wild wolf ancestors' adaptation to wintry conditions.

Unlike ground squirrels, which hibernate during the cold months, small rodents known as voles stay active year-round. They dig tunnels in the snow to feeding areas throughout the long Yellowstone winter. Voles weigh only a few ounces, but serve as a snack to a wolf, like a few peanuts to a person. Wolves, both pups and adults, seem to like the challenge of trying to catch them. Voles also function as live action toys for wolf pups.

I saw the lowest-ranking Druid female, 106, dig into deep snow, then shove her jaws into her hole. She must have heard a vole tunneling there. The wolf then pounced at a nearby spot and tried to pin down a vole with her front paw. Apparently she missed, for she dug at that site, then stuck her head into the excavation. Saddleback ran over and tried to help but just got in the way. The two wolves tussled for a few moments, then switched to playful wrestling.

Later in the month, I saw 106 playing with a live vole she had just caught. She pushed it around with a front paw, then dug it out when the vole burrowed into the snow. After catching it, she allowed it to run off but soon snatched it up again. 106 bedded down and gently jawed the vole. It jumped out of her mouth and ran off. She easily caught it again. By then the tiny animal seemed dead. One of the pups ran in and played with the lifeless vole. He let it roll down a steep snowfield, chased and grabbed it, then pushed the vole around with his nose. That motion made it appear to be alive again and when it slid down the snow, the pup chased and caught it.

I had never watched wolves during the early-winter mating season, so I looked for signs of romantic interest by 21 in the five females in his pack. Some people might assume that an alpha male would mate only with the alpha female, but because 21 had joined the pack and was not related to any of the females, he could potentially breed each one.

On January 5, I saw 42 go to 21 wagging her tail. She licked him on the face and he sniffed her rear end, which is the way male wolves check a female's breeding condition. He didn't do anything further, indicating she was not yet in season. Ten days after that, 21 went to 40 and tried to mount her. She was not quite ready to breed so she turned around and tried to nip him. 21 took the hint and walked off. I would see in the coming years that 21 had great respect for females and humbly accepted rejection.

On January 21, I was involved in my first wolf radio-collaring session. Collaring is done in the winter, partly because deep snow slows down wolves running from the helicopter and a slower target is easier to dart. Yellowstone's Wolf Project tried to have at least two collared wolves in every pack. An average of 25 percent of the park wolves had collars in a given year. The radio collars used by the Wolf Project weighed about 1.3 pounds, a touch above 1 percent of the weight of an average adult Yellowstone wolf. The batteries could last for up to four years, but sometimes failed earlier than that. When the collars failed, they might or might not be replaced, depending on how many wolves in the pack were collared, the importance of the individual wolf for ongoing studies, and whether the wolf could be captured again.

I spotted the Druid wolves in the early morning south of Soda Butte Creek, then called Doug Smith, who would be doing the darting from a helicopter. Kerry Murphy, the other Wolf Project biologist, was in a Supercub plane with pilot Roger Stradley. They would act as scouts to guide the helicopter pilot toward the wolves.

Around 10:00 a.m., the plane arrived, circled the pack, and keyed in on 40, the wolf the crew wanted to dart so they could replace her collar, which was no longer working. The helicopter came in from the west. Since the tracking plane often circles the wolves throughout the year, they usually don't react to its presence, but they run from the sound of a helicopter. They have learned that the sound of the rotors means there is about to be a darting operation.

Doug was securely strapped in and leaning out of the right side of his ship. He shot a tranquilizing dart into 40 and saw her run into some trees. The helicopter landed. He jumped out, followed her tracks, and found her lying in the snow. After giving her a physical exam for injuries, diseases, and overall health, Doug put a new collar around her neck. The crew radioed me her new transmitting frequency and I called back to them that I was getting the signal well on my receiver. The next morning, I saw her back with the pack, acting like nothing had happened.

On February 3, one of the Druids' lower-ranking females, 105, went to 21. He sniffed her rear end, then walked off, apparently uninterested. Soon after that, 42 approached him, wagged her tail, then moved it aside, exposing her rear end. Called tail aversion, it is a signal that the female is getting ready to breed. The next day, she did it again and he mounted her but slipped off. 21 wagged his tail, tried mounting her again, and once more dropped off. I soon learned that it usually takes several attempts by the male before a mating takes place.

His third attempt was successful and 21 and 42 got into a mating tie. No one has determined if there is any biological significance to the variations in the length of wolf mating ties, which can last from a few minutes to over a half hour. During the ties, the wolves seem to be locked together. If a disturbance occurs near them, they struggle to disengage, and that can take several seemingly painful attempts. 40, probably upset at the attention 21 was giving her sister, ran over and attacked 42, who was now on her back, still connected to 21. It was too late to prevent the mating, so 40 left and 42 got up and stood rear end to rear end with 21. The pair broke their tie at the six-minute mark. That was the first wolf mating I had ever seen, and it was especially meaningful since it involved 21 and 42. I had watched and studied them for years and felt they had an especially intimate relationship, probably because both of them had similar temperaments and seemed to prefer using cooperation over violence when dealing with other pack members.

A couple of days later, 21 sniffed the rear end of all the females in the pack except the alpha female, 40. Three days after that, he occasionally sniffed 40 but did nothing more. Later that day, 106, the lowest-ranking female, repeatedly stood in front of 21 and moved her tail to one side to indicate she was interested in mating. He mounted her, but soon slipped off. He kept trying and finally they got in a tie. During that process, 40 came over, interrupted them, and moved her tail aside for 21, but he ignored her, likely because she was not yet in season.

On February 12, after days of ignoring 40 and not even bothering to sniff her when she averted her tail, 21 showed more interest in her. 40 now upped her aggression to 42 and repeatedly pinned and bit her sister. 42 fought back, but always lost. Two wolf watchers who were a big help to me, Mark and Carol Rickman, noticed she had a bloody gash on her leg where 40 must have bitten her. I later saw 40 get up, walk over to 42, and bite her for no reason. I could see that 21 was paying more attention to 42 than to her, so 40 must have been aware of that as well and it likely contributed to her treatment of her sister. There was also the issue that the pack had five females who all could have pups. I figured that a wolf with 40's domineering personality would want to be the only female in the family to have a litter, and she was willing to use violence to achieve her goal. She had seen 21 breed 42 twice and likely suspected that her sister was pregnant.

Since I couldn't monitor the Druids every minute, there was a good possibility that 21 might mate with 40 when I was not watching. I did see him mount her several times, but never witnessed a completed tie. Four days later, 40 seemed to be out of season. She went to 21 and sniffed him. He ignored her and she did not avert her tail.

By late February, life in the Druid pack was getting back to normal. One evening I saw the two pups, Saddleback and Stripe, playing with each other. 21 joined in. He romped around his sons and wrestled with them, acting like they were all equals despite his much larger size and alpha male status. Later 21 lay down and one of the pups jumped on top of him. The young male put his front paws around his father's neck and they sparred with their jaws. One of the things that had always struck me about 21 was his playful personality. He was a big, tough-looking alpha male with heavy responsibilities, but he seemed to really enjoy roughhousing with his family.

Then 40 surprised me. She went over and gave one of the pups a submissive greeting and licked his face as a subordinate wolf would do to an alpha. She ran off, but soon stopped, turned around, and gave that pup a play bow before running off again, inviting the pup to chase her. The pup ran after her and jumped on her back. She playfully snapped at him, then ran off once more. I had never seen her act that playfully before with lower-ranking pack members. But then I realized that since both pups were males, they represented no threat to her status as alpha female.

On March 2, the Yellowstone Wolf Project late-winter study began. The project conducted two thirty-day studies each winter. The early-winter study ran from mid-November to mid-December and the late-winter study took place in the month of March. One of the tasks was for staff and volunteers to document all hunts and kills made during those thirty-day periods. I was on the Druid crew with a new volunteer named Melissa Andre.

On the fourth day of the study, we saw Saddleback and Stripe help the Druid alphas chase a mule deer doe. The wolves pulled her down before she could even start to run. She must have been very weak after a long, cold winter and a poor-quality diet of dried-out vegetation left over from the previous summer. The following day we checked on a big bison bull that for several days had been too incapacitated to get up. He died a few hours later. The Druids found the carcass and fed on it for many days. Older and weaker animals often die of natural causes in late winter, and their carcasses are an easy source of food for the wolves.

In winter wolves do not have to deal with grizzlies stealing their kills, for the bears are sleeping in their dens. That changes in early spring. I spotted the first grizzly out of his den on March 10. He looked fat and in great shape, despite not having eaten anything since late fall. The bear did look a bit groggy, though. I noticed that he stopped frequently to sniff the air, probably hoping to get the scent of a carcass. I had seen 42 go through that area recently. The bear found her scent trail and followed it, likely thinking it might lead him to a kill site and a free meal.

In mid-March, I got signals from the Druids at their forested den area north of the Footbridge parking lot. This was the place where 40 would likely have her pups. I could not see the wolves because of the trees, but I heard yelps of pain from the area. It was probably 40 attacking 42. A few minutes later, the pack came into sight and I saw 42 licking a fresh wound on her rear end. The two sisters were now close to five years old, about forty-two in human years. The pattern of 40 beating up 42 had been going on for years.

The next day, I saw 21 and his sons, Saddleback and Stripe, chasing a group of twenty bull elk. Saddleback ran past his father, caught up with the last bull in line, and grabbed one of his hind legs. The bull shook off the pup and ran. Both pups then ran along on either side of the bull. 21 was right behind them. 40 joined the chase and the bull ran behind a ridge. A minute later, a group of ravens repeatedly circled that area, indicating the wolves had killed the elk. The wolf tracking flight later that day confirmed that the pack had a carcass there. When that bull was later examined, he was found to have very poor bone marrow. Marrow tends to be the last of an elk's stored fat to be used, so the bull was in a very depleted condition.

Later 21 and the pups came back into view. Saddleback and Stripe started to play together, and their father joined in. I saw 21 wrestle with one pup, and the big alpha male let his son pin him. After that they made a game of nipping each other, then 21 chased the pup around the area. The other pup saw the game and joined in. The long winter was ending and soon the pack would be raising new pups, the most exciting time of the year for wolves.

2

Insurrection

ONCE WE GOT into April, I tried to keep track of where the Druid females would be denning. 40 claimed the main den near the Footbridge parking lot. But what about the other pregnant females? I had watched 21 breed 42 on February 4 and 106 a couple of days later. With a sixty-three-day gestation period, 42 was due April 7 and 106 soon after that. Although I did not see her and 21 get into a tie, 40 looked pregnant and she probably would have pups around the same time. On April 2, we got 42's signals from the south side of Lamar Valley, five miles west of the main Druid den. Three days later, I got signals from 106 to the northwest of Trout Lake, about four miles east of the main den. It looked like they were planning to have their pups at those sites.

By April 10, I was getting signals from both 42 and young female 105 from the western den. As 105 did not look pregnant, I assumed she was there helping 42 with her newborn pups. The battery on 103's collar was out so we couldn't track her, but from what I could tell, she was also helping 42. I had the impression that over the years, 42 had treated those younger females well, much better than 40 had done. 40 had also driven the mother of those females out of the pack. Those were likely reasons the two young females were helping 42 rather than 40. Stripe and Saddleback, now officially yearlings, were based at the main den site with 21 and 40. Tracking flights confirmed that there were three separate Druid dens that spring.

On April 13, I saw 21 leading a group of Druids toward 42's site. They disappeared into the forest where 105 was likely assisting 42 with her pups. I caught sight of 21 and other wolves romping around a small break in the trees. They seemed to be meeting up with the two females. When they all gathered around a large dead tree, we could see that 42 had dug out a den between the roots. Her pups were still too young to have ventured above ground. From then on, 103 and 105 were based at that site.

40 was in the visiting group and I noticed missing fur under her belly, a sign she was nursing. Thinking about how 40 had likely killed her sister's newborn pups in 1999 and 1998, I was concerned about the alpha female's intentions. I then heard yips of pain. I could not see the wolves among the trees but guessed that 40 was biting her sister. That was at 9:25 a.m. I continued to get signals from those wolves at 42's den throughout the day. In the early evening, 21 reappeared at the bottom of the forest accompanied by one of the yearlings. 40 and 103 soon joined them. I got signals indicating that 42 and 105 were back at the den. When it got dark, I reluctantly left the area and went home, worried that 40 might have harmed her sister or her pups.

The next morning, most of the family, including 103, 105, and the two male yearlings, was hanging out at 42's den and she was acting as though everything was normal. 40 and 106 were absent, likely at their own dens. The presence of so many pack members with 42 indicated that her pups were fine and the other wolves were helping her with them.

As the days went by, 21 repeatedly traveled from the main den to 42's den. One day he killed an elk by himself, then carried meat to the wolves at the western den. When he left, his belly looked far smaller, so he likely regurgitated to them as well. Adult wolves can carry far more meat internally than they can in their mouths, and a big male like 21 can gulp down twenty pounds of meat and bring it back to the den internally. The wolves at the den would lick his muzzle, triggering a regurgitation of what looks like pieces of stew meat. After dropping off all that food, 21 returned to the elk carcass, ate until he was stuffed again, then went to 40's den, carrying another big piece of meat. I got a glimpse of 40 the next day. Her coat was dirty, indicating that she had just come out of her den, and her nipples were distended, a sign she was nursing pups.

With pups at three separate dens, this was going to be a complicated denning season for the Druids. I felt that 21 would be up to the task, especially after I saw him cleverly deal with a problem involving food delivery in early May. He had crossed the road to the south to feed on a carcass and needed to bring back food to 40 and the pups at the main den. The wolf came toward the road, where a crowd had gathered to watch and photograph him. 21 veered off his route and traveled east behind a hill that put him out of sight of the crowd. Everyone rushed to their cars and drove east, hoping to get more shots of him. I was watching from a more distant area and saw 21 make a U-turn, trot west, then cross the road exactly where the people had just been. I recalled how the previous year, when he was trying to get back to the den, he had to walk six miles to the west to get around the line of cars following him. After he crossed the road, he walked another six miles to get back to the pups at the den. This time he had figured out how to outsmart the crowd.

I found a spot near the Yellowstone Institute where I had a good view of 42's den. 103, 105, and one of the yearlings were bedded down by the dead tree. The second yearling arrived and 103 ran to greet him. He regurgitated to her and she quickly ate the meat. Both yearlings went to the den entrance and looked in. 42 went to the spot where the yearling had regurgitated the meat and ate leftovers.

In the evening, 42 and a yearling went out hunting. They chased a cow elk and caught up with her. As 42 ran alongside her, the cow kicked out sideways and knocked the wolf down. After rolling over a few times, 42 jumped up and resumed the chase, seemingly unharmed. The yearling male grabbed a hind leg but was shaken off. 42 caught up with the elk and bit into a back leg. The cow shook that leg and the wolf lost her grip. I saw that the elk was running toward a deep section of the river. 42 caught up with her after she reached the water. The cow stopped and made her stand there. The yearling ran in and I got glimpses of the wolves trying to attack her. Then the second yearling joined them. I could see the wolves dashing back and forth in the water, but the riverbank blocked my view of the cow. I heard a lot of splashing. Then it got dark and I had to leave. The next morning I saw 42 and a yearling at the remains of that cow.

I was impressed by how much the other pack members—the two young adult females, both male yearlings, and 21—were helping 42. But I also thought about how 40 had likely killed 42's pups the previous two springs and worried she would try to do the same thing again.

After feeding, 42 headed up toward her den. 103, 105, and the yearling that had not gone to the carcass ran down to greet her. A minute later, I spotted a tiny black pup, about the size of a ground squirrel, by the den entrance. It was probably three weeks old. Two of the adults went up and sniffed it. Soon there were more pups stumbling around near the den, but the lighting was poor so I couldn't get a good count. Over the next few days, there were a lot of comings and goings as family members visited to check on the pups and on 42. She seemed to be comfortable allowing them to enter the den to see how the pups were doing.

One day I spotted a mountain lion uphill from the den site. If it found the den unattended, the lion could easily kill and eat the pups, but with 42, the two younger females, and the two male yearlings there, the pups were well protected. Plus 21 visited often and could handle that lion by himself.

On May 4, I saw three black pups at the den. 105 gently picked up one of them and carried it back into the entrance. The other two followed her inside. A fourth black pup was nearby. 105 was on babysitting duty that day, and her job was to make sure the vulnerable pups did not stray too far from the safety of the den. The next day, I saw what looked like a black wolf bedded down, then realized it was a pile of pups clustered together. There were at least five. I got a count of six the following day: five blacks and a light one. That was slightly higher than the average Yellowstone litter size of four to five pups.

The females made another kill that morning. 42 and 103 chased a herd of elk cows. 103 grabbed the slowest one by a hind leg while 42 ran in and bit into the cow's shoulder. 103 let go of her leg and grabbed the cow by the throat. The elk lifted the wolf off the ground, but she held on. 105 ran in and helped the other two wolves pull the cow down. That was the three-minute mark after the chase had begun. The cow was dead a minute later. The three females had co-ordinated their pursuit and attack as well as any Special Forces unit could have. They were an impressive team.

Later that morning, three wolves from a newly formed group known as the Tower pack came into the area. The founding members had once belonged to the Rose Creek pack, the family 21 had been born into. The trio moved toward the new elk carcass but backed off, probably when they got the scent of the Druid wolves. Soon 103 came down from the den and headed toward the site. She saw the three intruders and they spotted her. 103 continued toward them, possibly mistaking them for fellow Druids. Then she realized her mistake, tucked her tail between her legs, and ran back toward the den. All three Tower wolves ran after her. 103 disappeared into the forest below the den. The outsiders stopped and looked up the slope. A big male in the group sniffed around and probably got scents of other Druid wolves, including 21. The three wolves turned around, likely understanding they were outnumbered, went to the carcass and fed for a few minutes, then left the territory.

I was checking signals from 40's den several times each day. On May 7, I got signals from her and 21 at 8:47 p.m. that indicated they were leaving the main den and heading toward the Chalcedony Creek rendezvous site. That area had been used by the pack in past years as a convenient central location to bring the pups once they were capable of traveling and crossing roads, creeks, and rivers. I soon spotted the alpha pair meeting up with 42, 105, and Saddleback. 40 went after her sister and aggressively pinned her. 105 walked off and led the group toward 42's den. 42 squirmed out from under 40 and ran after 105. Then 40 raced over to 105 and pinned her. Her displays of dominance over the other females were more aggressive than usual.

Saddleback was now leading the group to the area below 42's den. 105 got away from 40 and ran after him. The alphas and 42 followed at a run. By 9:00 p.m., it was getting too dark to see, but the wolves were still heading for 42's den. I had a bad feeling about 40 going there and the theme song to the movie *Jaws* came to mind. Since two of the young females and both yearlings were spending most of their time with 42, it meant 40 was getting less food for herself and her pups at the main den. If she had killed her sister's pups once or twice before, she could do it again.

I got back to my observation point early the next morning. Signals indicated that 42 and 105 were near the den, and I could see 103 bedded down by the den tree. Everything seemed normal, but I couldn't see any pups. I tried for 40's signal. It was very loud, but I couldn't pin down the direction.

I wondered why she was still over in this area rather than at her den. After doing those checks on the Druids, I drove west to monitor the Rose Creek wolves.

At 7:40 a.m., I got a radio call from wolf watcher Anne Whitbeck. She told me there was a wounded wolf near the road north of 42's den. When I arrived, Law Enforcement Ranger Mike Ross was already on the scene. He took me to an angle where we could see the injured wolf hiding in a culvert a few feet from the road. The wolf was covered in blood and lying in three inches of cold water. It was 40. I was stunned to see her so helpless and fragile for she had always been in total command of whatever situation she was in. What had happened to her?

Mike and I went back to the lot to interview people and figure out what to do. We heard that 40 was first seen walking around near the road, then she ducked into the culvert, acting like she was looking for a place to hide. We looked back and saw that she was now up and moving around, but she was crouched over and obviously in pain. Blood was concentrated around her hips and at the back of her neck, the places other wolves usually attack, but since she was so close to the road, there was a chance the injuries had been caused by a vehicle. Normally the Park Service does not intervene if an animal has been injured due to natural causes. Because the wolf might have been damaged by human action, the rangers and the Wolf Project biologists decided to capture 40 and get her to a veterinarian for treatment. Kerry Murphy and Tom Zieber came out from the office with capture equipment, tranquilizing drugs, and a cage.

When everything was organized, we went out to 40, who now was lying on the ground. One of us touched her with a stick to test her reaction. She did not respond. We held her down, wrapped gauze around her jaws, then taped them closed. Her paws were also taped to prevent her from struggling and harming herself. I looked at her face and saw that her eyes were open, but she did not seem fully conscious. She must have been in shock and was very wet and cold. I got the sense that she was not suffering.

Once she was restrained, we took a closer look at her wounds. There were bite marks on her rear end and belly. But far more serious was the gaping wound on the back of her neck. I could see the white of her spine through that deep gash. That settled the issue: she had been attacked by other wolves. This was a nursing mother with very young pups at her den. If we abandoned our plan to get her medical treatment, not only would she die, her pups might starve to death.

I looked at the crowd of people watching us. Most of them knew the Druid wolves and were in great distress when they saw 40's injuries. All of us with 40 agreed that we would stick to the original plan of getting her to a vet. We wrapped her up in a blanket, carried her to the lot, put her in the back seat of a crew cab pickup, and drove the half mile to the Lamar Ranger Station, where Mike Ross lived. Once we got there, he filled several jugs with hot water and put them around the wolf to warm her up. Tom Zieber and I stayed with 40 in the truck, with the heater on full blast. I saw Tom turn to look at her. After a few moments he quietly said, "She's stopped breathing." Mike was an emergency medical technician and he confirmed that 40 had passed away.

We sent 40's body to Montana Fish, Wildlife and Parks for a full autopsy, and they told us her death was caused by bites from wolves. On the morning we found her, there were no signals from any wolves other than the Druids. That meant her own pack had killed her.

I tried to imagine what must have happened. I think that 40 went to 42's den, saw one of her pups, and went after it, intending to kill that pup and then the rest of the litter. At that moment, I think 42 had finally had enough of her sister's violence and attacked her, trying to save the pup. But in a fight, 42 would not have the aggressiveness or strength to defeat 40.

40 appeared to have been bitten by more than one wolf. That was the clue to figuring out what had likely happened. As the two sisters fought and 40 was winning, 42's allies, young females 103 and 105, must have jumped in on her side, making it three against one. All the unnecessary aggression and attacks 40 had imposed on those females over the years came back on her. They must have defeated her, beaten her up, but then stepped back and let her go. That was how she got from the den to the road where we found her.

40 and 42 had lived together every day for over five years. 40 used aggression to deal with issues while her sister used cooperation. 42 had suffered from 40's bullying her whole life. Her longtime cooperative relationships with the younger females saved her on that day. 42 changed the equation of her life from thinking that she could not stand up to her oppressor to confronting 40 with the help of her allies.

I went to the main den and got 21's signal there. That meant he was trying to care for 40's pups. At this stage of their lives, the pups would still be relying on their mother's milk. 21's great strength, hunting prowess, and fighting ability were useless at that moment. I got a signal from 106 near her den site. I later spotted her moving toward the main den. She was apparently not involved in the attack on 40. I could see that 106 had distended nipples, proof that she was nursing pups. I lost her entering the trees around the main den. When she met up with 21, he would likely take her to 40's pups. Would she nurse them or walk away, remembering the abuse she had endured from their mother? I did another check on 21 and his faint signal indicated he was down in the den. Since nothing was happening, I headed home for a break, my head reeling from the events of the morning. The Druid pack was never going to be the same.

In late afternoon, I went back to the main den and got a very weak signal from 21. Three minutes later, his signal got much louder, meaning he was coming out of the den. Then I went to the western den and got signals from 42 and 105. In addition to my concern about 40's litter, I had been worrying about 42's pups. Had 40 killed some of them before the other females attacked her? I put my scope on the den tree and soon saw 103, 105, two black pups, and a light pup. A wolf watcher told me she had just seen at least six pups there. I could now relax a bit. It meant that 42 and her female allies had saved the pups from 40.

The next morning, May 9, the Druids had a fresh kill just north of 42's den. When I arrived, 21 and Saddleback were feeding, and 42 and 105 were heading back toward 42's den. The four-week-old pups would still be nursing and normally would not start to eat meat for another week or so. 21 left the carcass with a full belly and meat in his mouth, headed toward the main den. As 40's pups likely were also about a month old, they might try to nibble on meat brought in by their father, but they desperately needed milk. I pictured the pups crying out from hunger and 21 standing next to them, feeling helpless.

Two days later, I got signals from 21 and 42 at 42's den in the early morning. That evening, the two signals came from the trees surrounding 40's den. I got word that both wolves had been seen going up there an hour earlier. It appeared that 21 had brought 42 to the main den to help him with 40's pups. But then I thought of something much darker. Since 42 had suffered so much from her sister, would she be better off if she killed 40's pups? That would make sure none of them grew up with their mother's super-aggressive personality and ensure more family support would go to her pups. Later I got a loud signal from 21, but 42's was very weak. That meant that she was likely in the den with her sister's pups. What was she doing: nursing them or killing them?

Early the following morning, May 12, I didn't get 42's signal at the main den. I drove west and spotted her heading back toward her own den. As she trotted up through the trees, 103 ran downhill to greet her. A minute or two later, 42 was at the entrance to her den nursing a big cluster of pups under her belly. They were too close to each other to count. The pups went back into the den and I soon lost sight of their mother.

Sixteen minutes later, I heard that 42 was traveling east with a pup in her mouth. I arrived just in time to see her rapidly approaching the Lamar River. The pup in her jaws was limp and seemingly comfortable with being carried. I climbed a hill to get a better view and saw 42 wade into the fast-flowing water. She began swimming when she reached deep water, holding her head high to keep her pup from drowning. When she arrived at the far bank, she headed toward the section of the road where the wolves crossed to get to the main den. Several cars stopped to watch her and 42 had to go farther east before she could bypass the drivers who were following her.

A few minutes later, I did a signal check and got 21, 42, and 105 at the main den. Twelve minutes after that, I no longer got signals from 42. That meant she had taken her pup into 40's den. Based on the signals I was getting, she was underground for the next fifty minutes. In late morning, we saw 42 and 105 heading back to the western den. I lost them going up the forested slope at 11:07 a.m., about five hours after 42 had left the site with the black pup. When 42 arrived at her den, she looked in the entrance and at least one pup greeted her.

Four minutes after that, 42 and 105 came out of the trees below the den and both females had black pups in their mouths. 42 led east on her earlier route and I saw her swim the river and cross the road, then go up to the main den with her pup. I had lost track of 105 near the river crossing. I went back and saw 105 carrying her pup back toward the western den. She was probably afraid of crossing the flooded river with the pup. Not knowing what else to do, she took it back home. I saw three other black pups at 42's western den. The other young female, 103, was with them.

In early afternoon, Tom Zieber saw 21 and 42 come down from the main den, go to a recent carcass, and feed.

Both wolves then went up to the western den and 42 soon reappeared with another black pup in her mouth. She carried it up to the main den at 4:15 p.m. It was at least the third pup she had moved there.

Early the next morning, I got 21 and 42 at the main den, along with 105. In late morning, 42 returned to her own den and came back out with another black pup in her mouth. She took it on the usual route to the main den. That was the fourth black pup she was seen carrying there, and during the night she probably brought more to the site. It appeared that 42 was planning to move all her pups to the main den. That evening, I got signals from 21, 42, 105, and 106 at the main den. The other young female, 103, was seen heading that way from the direction of 42's den.

We tried to figure out what was happening. The biggest issue was 40's pups. What had 42 and the other females done about them? Were they caring for them or had they killed them? With 106's signal now coming from the main den, it appeared she had also moved her litter there. It would be safe for them now that 40 was dead. The Druids soon stopped going to the western den, indicating that all the pups had been moved from there to the main site.

Now the Druid pack was much better organized. All the surviving pups were together at the main den so adults coming back with food from a kill only had to visit one den site to feed them. But we still didn't know how many pups were now at the main den or what had happened to 40's pups. We knew 42 had six pups, and we knew both 106 and 40 had given birth to pups, but we didn't know what size litters they each had. And now all the pups were mixed together in one den.

I then saw something that made things even more interesting. I spotted 105 traveling and noticed that she had distended nipples. That was a sign she had given birth to pups and was now nursing them. 21 must have bred 105 when we were not around. She had been based at 42's den, so some of the pups we had seen there must have been hers. I had thought she was acting as a babysitter when actually she had been sharing 42's den and using it for her own pups. That would have given her a strong motivation to help 42 protect both litters from 40.

On May 16, we saw a significant open wound on 42's hip. I had often seen 40 bite her at that spot and realized that the injury had probably happened when 40 went to her sister's den. That supported the theory that the two sisters had fought over 42's pups.

That same day, I noticed that 105, the highest-ranking of the three young females, did not raise her tail when walking by 106, the lowest-ranking female. Aggression among the four adult females was at a low ebb. They were working so hard to feed the pups that they were putting aside any squabbling. I also thought that 42's cooperative, non-aggressive character was influencing the other females. She had shared her den with 105, a wolf that was subordinate to her, then consolidated four litters from three separate sites at the pack's traditional den site.

Something significant happened later that day. For the first time, I saw 42 do a flexed-leg urination, which is usually done only by alpha females. 21 immediately went to her site and marked it. That double scent marking confirmed that 21 and 42 were now the Druid alpha pair and she was the pack's new queen. The pair went to a new carcass and she fed right away. 21 bedded down and watched her, letting a nursing female have preference over him at a feeding site. I wondered if he was thinking about how much more tranquil family life was going to be, now that 40 was gone and 42 was running the pack.

Soon after that, I saw 106 walk up to 42 with her tail held in a neutral posture. She sniffed and greeted the new alpha female. 42's tail was also neutral, and both wolves seemed relaxed and totally comfortable with each other. That interaction was the opposite of how 40 treated lower-ranking females and one more sign that 42 had instituted a new era in the pack, one that was more supportive and respectful. An earlier analysis of 106's DNA concluded that she had been born to 41, sister to 42, but a later, more advanced test showed that 42 was her mother. That meant that 106's pups were 42's grandchildren.

Despite 42's consolidation of the pups, the Druids had a tough road ahead of them. There were four litters now based at the main den. 21, 42, and the other five adults would have a huge task in the coming months as they went out on hunts and tried to bring back sufficient food for the large family. The heaviest responsibility would be on 21 as alpha male.

3

Counting Pups

AS WE GOT further into spring, elk cows in Lamar Valley were getting ready to have their calves. I saw the first one on May 22. That meant additional hunting opportunities for the wolves at just the time when their pups were switching over to eating meat.

I left my cabin at 5:07 a.m. on May 24 and drove west toward the Druid den area. I was half a mile away when I saw Wayne Kendall, a former police chief who had a second home in Silver Gate. From his expression, I could see that something was wrong. Then I saw a gray wolf lying on the road near him. It was the male yearling Stripe. He must have run across the road during the night and been hit by a vehicle. Brain Chan, the district ranger, arrived and took the wolf for an autopsy. That brought the adults in the Druids down to six. With so many mouths to feed, it was not a good time for 21 to lose a pack member.

In early June, I watched the Druids set out on a hunt. 106 spotted an elk cow and moved toward her. The cow chased her away, but the wolf circled back and sniffed around where the elk had first been. 21 and 42 joined her in searching the area for a newborn. 106 suddenly rushed ahead and put her head down. She had found the calf. The alphas ran over and the three wolves got the calf. 21 and 42 carried it off, but then 21 walked away and let 42 and 106 have it to themselves. I was very impressed by his behavior. He had unselfishly passed up a meal so that the two nursing mothers could eat.

That summer, Tom Zieber and I were both working for the Wolf Project in Lamar Valley. We kept records on wolf behavior, helped visitors see the wolves, gave talks to the crowds in the valley, and tried to minimize behavior by visitors that might negatively impact the wolves. The most common issue involved road crossings. When wolves were approaching the road to leave the den on a hunt or to return to feed the pups, people would often drive to the likely crossing spot for photographs and unintentionally block the wolves.

An example of that took place on June 3. When 21 was spotted carrying an elk calf back to the den, many drivers ran to their cars, intending to head down to his crossing site. I put on my orange safety vest, grabbed the red stop sign the law enforcement rangers had lent me, and held up traffic until he made it across. I felt like a school crossing guard helping kids cross the road safely. If Tom was on duty with me, he would stop traffic on one side of the crossing area and I would do the same on the other. When Park Service naturalist Bill Wengeler was working with us, I would often hike up a slope south of the den called Dead Puppy Hill. The area got its name in 1995 when the Crystal Creek wolves killed some coyote pups there. I would call down to Tom and Bill on the park radio to let them know when wolves were heading toward the road, and they would work together to stop traffic.

On June 7, I left the park for a few days because my mother died. I got back to Lamar on June 11 and saw some of the Druid adults that evening. The next morning, I started a long run of going out before sunrise every day to look for wolves and study them, a streak that would last for years.

In mid-June, I went up on Dead Puppy Hill to watch the den. I soon saw 21 come out of the trees with a big group of pups. It was hard to get an accurate count as they ran back and forth, but I eventually saw nine blacks and seven grays. Sixteen pups were a lot for a wolf pack with only six adults, but all of them looked fat and healthy. We had no way of knowing which pups were born to which mother. A pup born to one female might be nursed by another mother wolf, so even if we witnessed nursing it would not determine maternity.

A few days later, I walked south from the Footbridge parking lot and saw pups up at the den forest. I rushed back to the lot and got the people there to come back with me, and all sixty of them got to see pups through my spotting scope. I figured that every time I helped someone see wolves, it would be one more person on the side of wolves.

In late June, I was on Dead Puppy Hill with sixteen park visitors. A grizzly came into sight about seventy-five yards away. I told everyone to yell at it to scare it away, but the bear continued toward us. When it got to within forty yards, it sat up on its rear end and stared at us. Most of the people laughed at the sight. That laughter worked better than the yelling, for the grizzly got up and walked away.

I had lived in grizzly country for twenty-five years by that time. That incident reminded me of another close call in Alaska at Denali National Park. I was riding my bike up a steep section of the park road. It was early in the morning and no other people or vehicles were around. As I reached the crest of the hill, I looked in my mirror and saw a grizzly charging at me. I took a moment to consider what might happen if I continued on through a long, steep downhill section of the road. I could go faster than the bear down the slope, but it would catch me at the bottom of that hill. So I stopped, put the bike between me and the charging grizzly, took out my bear spray, and removed the safety cap. That startled the bear. It stopped, looked at me for a few moments, then turned around and walked off.

One evening in late June, I saw four Druid adults come down from the den, cross the road, and continue south. Two minutes later, seven pups ran down to the road. But instead of following the adults, the pups sniffed around the pavement and played there. Tom Zieber was in the next lot to the west. He stopped traffic there, while I held cars to the east. We assumed the pups would run back up to the den within a few minutes, but they were fascinated with the road and stayed put.

I called for help and Eric Barron, a patrol ranger, joined us. The pups had been on the road for thirty minutes by then and some drivers were getting upset at the delay. It wasn't safe for those pups to be so comfortable with being on the road, so Tom, Eric, and I worked out a plan to get them off.

I got in Eric's patrol car and we drove toward the pups with his multiple flashing lights on. The pups saw us coming but did not seem concerned. Eric stopped close to the pups and they still did not get off the road. I got out and yelled at them. That worked, and all of them ran back uphill.

The following day, 21 came down from the den and crossed the road. Once again, seven pups followed him and stopped on the pavement. Tom drove toward them and the pups ran back uphill as soon as they saw him coming. That indicated that the pups were learning that the road could be a scary place. I thought about how Stripe had been killed when he tried to cross the road and was glad these pups were learning about road safety.

In early July, a grizzly took over a Druid carcass and kept two of the young females and Saddleback from feeding on it. When 21 arrived on the scene, he ran at the bear. The other three wolves followed his example and the grizzly ran off. 21 stopped twenty feet past the site, then turned around and fed along with his packmates. It was impressive to see how decisive and effective 21 was in dealing with that bear. He knew what to do and did it. The other wolves trusted his leadership and backed him up.

After feeding for thirty-five minutes, 21 left and went to the den. I saw at least fifteen pups running to meet him. They leaped up to lick his mouth and he regurgitated a huge pile of meat to them. Then Saddleback arrived. The pups raced to him hoping for another feeding. More pups joined the mob. They all ran through a meadow and I counted twenty-one pups: thirteen blacks and eight grays! That was an extraordinarily high number. I had never seen anything like it before.

Since Yellowstone litters average four to five, twenty-one pups was evidence that there were four litters at the den and that 42 was raising 40's pups, despite everything the former alpha female had put her through. All those pups must have been sired by 21.

How could just six adult wolves provide enough food to keep all twenty-one pups alive and healthy? It was a daunting task. For the past two years, the Druids had had poor pup survival. Only two of the six pups born the previous spring had survived to the end of the year, a survival rate of 33 percent. If that same rate occurred this year, just seven of the twenty-one pups would make it to year's end, a dismal prospect. Last year 40 ran the pack. This year 42 was the alpha female. The fate of those pups would depend on her leadership and organizational abilities.

I began to see the pups hunting for voles in the marsh near the den forest on July 5. That was a good sign. If the pups could acquire some food on their own, it would ease the burden on the adults bringing in food. As in past springs, I saw pups doing play bows to each other, an invitation to play games. They wrestled, chased each other, set up ambushes, and played tug-of-war.

One morning, I saw a grizzly walk up into the den area and head toward some pups. 21 was south of the road. He must have gotten the grizzly's scent for he ran north, crossed the road, and rushed up the slope. Tom was on Dead Puppy Hill and he radioed me that 21 charged at the grizzly and bit it. This was a dangerous move since the bear could have killed him with one bite or a swat from a paw, but 21 was fearless when his family was in danger. I later caught glimpses of the bear close to some of the pups, but it did not seem interested in them. Perhaps it was there looking for an elk carcass it could steal from the wolves. Regardless of the grizzly's intent, if a naive pup got too close, it would take just one crushing snap of its powerful jaws and the pup would be dead.

That bear was a day too early. The following day we saw that the Druids had killed a big bull elk just east of the marsh. I watched pups go to the site and feed. One of the three-month-old pups dug a hole and buried some of the meat for later use. I felt the pup did that by instinct, since it probably had never been at a fresh carcass before.

The pups occasionally followed adults down to the road, but now knew that the road was not a safe place to hang out. One day six pups trailed 21 to the road and ran back uphill when they saw a car moving toward them. That was exactly what we wanted them to do. Later, when the pups were older, the adults would need to get them across that road and bring them to the Chalcedony Creek rendezvous site on the other side of the Lamar River, but right now it was good that they were afraid of it.

21 was the pack's workhorse when it came to bringing food to the pups. In mid-July, he returned to the den from a carcass and regurgitated to them five times, then left the den within the hour to go back for more meat. He was working so hard to hunt, make kills, and bring meat to the four litters that he had to be exhausted. One morning, I saw six pups following him around the den area, pestering him for a play session. He ran down to the road, quickly crossed to the south, and bedded down in a thick willow patch a hundred yards farther south. It looked to me like he was hiding

so he could get some sleep. Those pups, wary of the road, turned back. After napping for thirty-six minutes, 21 got up, recrossed the road, and went back up to the pups.

Early on July 15, I saw some of the Druid adults and ten pups at a carcass south of the road. The adults must have gotten the pups across during the night when there was no traffic. Later 42 and other adults got those pups to cross the river and took them to the rendezvous site at Chalcedony Creek, which was an ideal central location to stash the pups while the older wolves went on hunts. Having the pups there also greatly reduced the number of times the adult wolves had to cross the park road. The following day all twenty-one pups were at the rendezvous site.

Getting all those pups across the river must have been a massive task. I had previously seen how 42 was a master at coaxing pups into the water and getting them to swim to the other side. If some hesitated, she would grab a stick, show it to those pups, then rush into the water. They would always run after her, thinking this was a game, and be swimming before they realized what they were doing. In later years, I saw that 42 was the one who always organized the mission of getting the pups to that rendezvous site. She was gifted in knowing how to plan and execute a complex operation. If 42 had been born a woman, she could have been an extremely effective general or US president.

One evening, while I was watching adults and pups at the rendezvous site from a hill north of the road, I saw three people cross the river and head toward the pups. Remembering an incident in 1998 when the Druids abandoned their site after hikers got too close, I tried to intercept the people.

They ended up turning around, so I went back uphill and continued to monitor the wolves.

The next morning, July 17, we saw two pups at the rendezvous site, but no adults. I wondered if the adult wolves had seen those people and led the other nineteen pups away. I went up on Dead Puppy Hill to see if there were pups in the den area but found none. 21 came back to the rendezvous site early the next day with meat in his mouth. He gave it to the two pups, then regurgitated twice to them. That evening he fed them two more times. He stayed with them through the next day, then left. Kerry Murphy did a flight on the twentieth and later told me that he saw 105 and fourteen pups on Mirror Plateau, an area south of Specimen Ridge, the ridge that forms the southern boundary of Lamar Valley. 21, 42, and 106 were hunting east of there.

The two pups at the rendezvous site seemed all right with being alone over the next few days. One day a grizzly came within thirty yards of them. The bear casually looked at them and both pups took a few steps toward the grizzly, then sat side by side and watched as it walked by. Later a herd of bison passed by the pups. They howled often, probably trying to contact the adults, but got no response. They entertained themselves by wrestling and playing together.

A researcher told me that vole numbers were high that summer and many of the little rodents were having a second litter. Grasshoppers were also present in large numbers. I saw a pup snap at something flying by and then chew whatever she caught. Right after that she reached down and snatched up a vole. The high density of voles and grasshoppers ensured these two pups had something to eat even when the adults were not around. Those abundant sources of food could be a decisive factor in pup survival that year.

After five days, 105 showed up and stayed with the pups for a few days. All three wolves were gone on the thirtieth, so she must have taken the pups to join the rest of the family.

In early August, the Druids and their pups were based at their Opal Creek rendezvous site, about five and a half miles uphill from the site at Chalcedony Creek. That meadow on Specimen Ridge was close to where elk spent the summer, attracted by the lush grasses growing in the high country. The Druids had several routes they took up and down the ridge when they traveled from Opal to the floor of the valley and back again.

A week later, the family was back at their Chalcedony rendezvous site with seventeen pups. When 21 arrived, the pups mobbed him, and he regurgitated a big pile of meat to them. After they fed, the pups bedded down with their father. Later two more pups showed up. There were eleven blacks and eight grays. Two black pups were missing.

There was a major development on August 15. An uncollared gray male we didn't recognize was at the rendezvous site with the Druids. He seemed to be totally accepted by 21, the other adults, and the pups. We never figured out who he was, but I suspected that he was a Rose Creek wolf and therefore a relative of 21, perhaps a younger brother. The Druids needed all the help they could get for their pups, so this recruit, whom we called New Gray, proved to be a major asset to the pack. That day New Gray regurgitated to a pup, then had a friendly greeting with 105. Later he did the same with 103 and did two more regurgitations to pups.

21 returned to the rendezvous site and also fed the pups. A twelfth black pup turned up, so now we were short only one black. That missing pup never showed up. Bears, coyotes, and eagles could kill wolf pups, and sometimes wandering pups just got separated from the adults.

A few days later, a grizzly approached the pups at the rendezvous site. New Gray was the first wolf to see the bear. He charged and drove the grizzly off. 21 and 42 stood up, but just watched the interaction. Having this extra male to guard the pups enabled the alpha pair to get more rest.

The Druids left their Chalcedony rendezvous site early on August 21. Doug Smith flew two days later and found them south of Cache Creek, a tributary of the Lamar River. We waited two days, then Bill Wengeler; Ray Rathmell, the Pebble Creek Campground host; and I hiked up the Lamar River Trail to the Cache Creek Trail intersection to see if we could locate the wolves. Backpackers there told us that they had heard howling in the area the previous night. We walked up the creek and soon found wolf tracks. I got good signals from 21, 42, and 105 from Death Gulch, a barren gully on the south side of Cache Creek. Most of the trees and bushes there were dead, and a strong sulfur smell from a hot spring bubbling up from the small creek in the gulch permeated the air. In past decades, gases from the hot springs in the area had killed many elk and bison in the gulch, along with the plants, but the geothermal activity had declined over the years and those gases were no longer at a fatal level. The temperature of the water was not high enough to harm wildlife.

An hour later, we spotted 42 and six pups nearby, then lost them over a ridge. Seeing the Druids so close to Death

Gulch was like watching a scene in a postapocalyptic movie where only wolves had survived a disaster that has killed off all other forms of life. If they made that film, 21 could have played the Mad Max character.

The pack was back at their Chalcedony rendezvous site in late August. I counted all twenty pups and seven adults, for a total of twenty-seven wolves. New Gray did a raised-leg urination and both 21 and Saddleback marked over his site. The scent marking by the younger male showed that he was maturing.

Late one day, I watched 21 with the setting sun behind him. His shadow projected fifty feet toward me. The image struck me as symbolic of the impact he was having on Yellowstone and on the people that knew him. I had begun to think we had entered a golden age for wolves in the park. It was a time of legends, a time when giants strode the land. It was the time of 21 and 42.

4

The Pups Grow Up

By mid-September, the twenty pups, now five months old, were at an age where they were likely to survive their first year. I noticed that the alpha pair interacted with each other much more now that the pups needed less attention. 21 in particular seemed more relaxed and he played with the pups frequently. Even though the pups were big enough to feed themselves at kills, 21's paternal instinct was still strong, and he carried meat to them and did regurgitations. The entire family often traveled in a single-file formation, an impressive sight. 21 or 42 usually led the pack, but not always. During one trip, Saddleback, the youngest adult, was out in front, 21 was fourteenth in line, and 42 was twenty-fifth. Now that the pups were doing so well, the alpha pair could take it a bit easier and let younger adults get experience in leadership.

Grizzlies were less of a threat to the pups now. I saw a grizzly chase a black pup, and the young wolf easily outran the bear. Then the pup fearlessly circled the grizzly and seemed to be daring him to chase it again. Other pups saw that interaction and ran over. Realizing he was outnumbered, the bear moved off and nineteen pups followed. When he turned and charged at them, the pups scattered, then came right back and surrounded him. Soon they grew bored with that game and left the grizzly alone. The interaction was an important lesson for the pups. They learned that they could outrun a grizzly.

Late in the month, I saw the Rose Creek wolves in their territory at Slough Creek. Their founding alpha female, wolf 9, mother of 21, had been driven out of the family by her daughter 18. Later 9 started a new group east of the park, known as the Beartooth pack. It was very successful and continued to exist long after 9 passed away. Other members of the Rose Creek pack had left to form the Tower pack. One day I watched the Rose Creek wolves run into a ravine. A few moments later, I saw a mountain lion racing out of that area with the pack right behind it. It streaked toward a nearby tree and leaped up high on the trunk. The wolves ran to the base of the tree and looked at the lion, then turned around and rushed back to the ravine where they had first encountered the cat. I saw them feed on a carcass there that likely had been killed by the lion. The wolves ran back to the tree several times and looked up at the lion, who was calmly sitting on a big branch. At one point 18 barked at it, just like a dog barking at a house cat in a tree.

Despite being a five-year-old, middle-aged alpha female with a lot of responsibly, 18, like her brother 21, still could be playful. After the lion incident, I saw her bounding after a vole. She did a leap and pounce and got it. A moment later 18 put it down, let it go, and pushed it around with her nose. When it ran off, she caught it again.

In October most of the Druids made a brief foray into the Rose Creek territory at Slough Creek. They might have been searching for prey or testing the territorial boundary between the packs, but either way the Druids would have known they were in country claimed by a neighboring pack. After a while the Druids moved south toward the park road, likely intending to join 106 and the pups, who had stayed on that side of the road. The main group approached the road but turned back when they saw people stopping and getting out of their cars. The vehicles and people were between the two groups of wolves. Tom Zieber and I switched from our wolf researcher roles to our school crossing guard personas and stopped traffic. Some of the pups in the northern group crossed through that gap in traffic, then 21 and the rest of the wolves followed.

A few days later, I found nineteen of the Druids back at Slough Creek in Rose Creek territory. 21 and 42 were not with them and I had not gotten their signals anywhere that day. That was unusual, for the alphas were normally with the main Druid group. Roger Stradley, the pilot who flew the tracking plane, called me later that day to say he saw 21 and 42 by themselves way up the Lamar River. Maybe they were taking a break from the twenty pups. After all the drama and stress of the spring and summer, I felt they deserved some time to themselves.

I was constantly learning more about cooperative behavior among wolves as I watched the Druids. After the alphas rejoined the pack, I saw 21 feeding on a carcass, then walk off with a piece of meat. 103 went to him and he gave the meat to her. Then a gray pup ran to her and she passed the meat on to it. That sequence of events showed how sharing families of wolves are with each other.

On October 9, I noticed a change in the male hierarchy. New Gray had been subordinate to Saddleback, but on that day, he was clearly dominant to the yearling. 21 stayed out of whatever was going on between them. A few days later, 21 and New Gray were having a playful tug-of-war over an elk antler, which showed that they were getting along well. Three pups ran over and one of them stole the antler from the two big males.

21 and 42 were often affectionate with each other. I saw 21 lick 42's face for a long time as he wagged his tail at her. Later she did teasing feints at him, then they gently wrestled. After that he did a play bow to her, then they romped around together. They had known each other for three years, but for most of that time, 42 had been bullied and dominated by 40. Now that 40 was gone, the two wolves seemed carefree and acted like a pair in their early courtship days.

As we transitioned into fall, the elk that had summered in the high country began migrating back to their winter range in Lamar Valley. In the cooler temperatures, the Druids became even more playful with each other. One November day, New Gray and a group of pups chased 103. Off to the side, two gray pups played together. One grabbed the other and dragged it across a snowfield. Other pups played in pairs and subgroups. Then thirteen of the pups came together and played. 21 had been bedded down with 42, but he could not resist joining the pups. He went to a black pup and romped around with it. They played chasing and ambush games. When they were done, 21 went back to 42, and the alpha pair romped around together. I couldn't think of a scientific term to describe their mood. The only word that seemed to fit was joyful.

On November 23, all twenty-seven of the Druids were back in Rose Creek territory at Slough Creek. Doug flew over and called down to let me know that four Druid wolves were following the scent trail of a Rose female and two black pups. I rushed up on a hill and got in place just in time to see 42 pinning one of those black pups. She must have just caught it. Then 42 stepped back and let the pup go. It jumped up and ran off. That pup joined the other Rose pup, and both ran away. 42 did not chase them. Her restrained behavior was consistent with what I had seen of her over the years. 42 saw rival wolves near her own pups, chased and caught one that turned out to be a pup, and let it go.

I went up on South Butte on December 6 to view a section of the park known as Blacktail Plateau, the home base of the Leopold pack. Their territory was just west of where the Rose Creek wolves lived. I saw twelve Leopold pack members bedded down. Alpha male 2, who was five and a half years old, had a lot of gray streaks on his black coat. He got up and went to 7, his longtime mate. The two met as yearlings and had been a pair for four years. I recalled how in the spring of 1996, in the early months of their relationship, I had seen a lot of affectionate interaction between the two. Their relationship seemed especially close, like the one between 21 and 42. When the male reached 7, she jumped up and wagged her tail, then did playful feints at him. He wagged his tail at her. She dropped into a play bow, romped around, gently nipped him, and jumped up in the air several times. All that demonstrated the emotional bond that still strongly connected them.

In late December, seven Druid pups chased three bull elk through snow. The lead pups tripped when they hit deep drifts and the bulls got away. On December 30, I spotted the pack chasing a cow elk across a snowfield at Slough Creek. Saddleback led the pursuit and other wolves ran single file behind him in his broken trail. The yearling reached the cow and grabbed a hind leg. That slowed her down. A black ran past Saddleback and got a holding bite on the elk's neck. More Druids ran in and they brought her to the ground.

I counted seventeen wolves near the kill site. Six adults and three pups were nearby and were not involved in the hunt. That meant the kill had been made by Saddleback and sixteen pups, all of them sons and daughters of 21. The black that grabbed the cow's neck had to be a pup, an impressive accomplishment for an eight-month-old wolf. That moment was an important milestone for the family as it showed the pups could now help the seven adults hunt and make kills.

Wolves usually have an advantage over elk on snow-covered ground. When wolves run, their paws spread out as they touch the snow. That extra surface area creates a snow-shoe effect, a big advantage when they are chasing elk. Adult male and female wolves in Yellowstone average about 100 pounds with big males occasionally going up to 135. Adult elk can be 500 to 700 pounds, so their extra weight often slows them down, especially when they have to slog through deep snow. Wolves conserve energy by traveling single file through snow. The lead wolf bears the brunt of breaking trail, but other pack members will trade off in the lead position and do their share of that exhausting work. Old pack members often stay at the rear of the group and benefit from the efforts of the younger lead wolves.

As the year ended, I thought about what a tremendous accomplishment the Druid adults had achieved with pup survival. Twenty of their twenty-one pups made it to the new year, a 95 percent success rate, compared with 33 percent the previous year. There was one major difference between those two years. This year 42 was the leader of the pack, rather than 40. With her benign, non-aggressive personality, 42 seemed to get all the wolves to cooperate and work together. I later watched a 2014 documentary on the PBS NOVA series titled *Inside Animal Minds*. They interviewed animal researcher Brian Hare, a professor of evolutionary biology at Duke University, and he spoke of matters that applied directly to 42:

> There are other animals that experience things that we also experience. Challenges that we face every day as social animals have a big role in the evolution of bigger and smarter brains because in certain situations the creature that can cleverly negotiate or extend a helping hand is often the one with the best chance of survival. We often think the biggest, strongest, most competitive individual is going to be the one who will survive and reproduce, but we see again and again in evolution that's not the case at all. At times what is going to be favored are things that lead to

> better cooperation so that you can work together to solve problems that otherwise you can't solve on your own. That requires tolerance, that sometimes requires a lack of dominance. It's not always the big guy that wins.

As I have watched and studied wolves for over forty years, I am continually impressed by how much they cooperate. Wolves have evolved to work together as a team to hunt prey animals considerably larger than they are, raise pups, and defend their territory from rival packs in much the same way early humans living in small groups shared duties such as hunting, food gathering, child rearing, and defense of their land. That suggests that wolves and humans evolved along parallel lines. What worked well for wolf packs also worked for bands of early humans. Native American people understood that. They watched and studied wolves and applied what they learned from them to their own lives.

PART II

2001

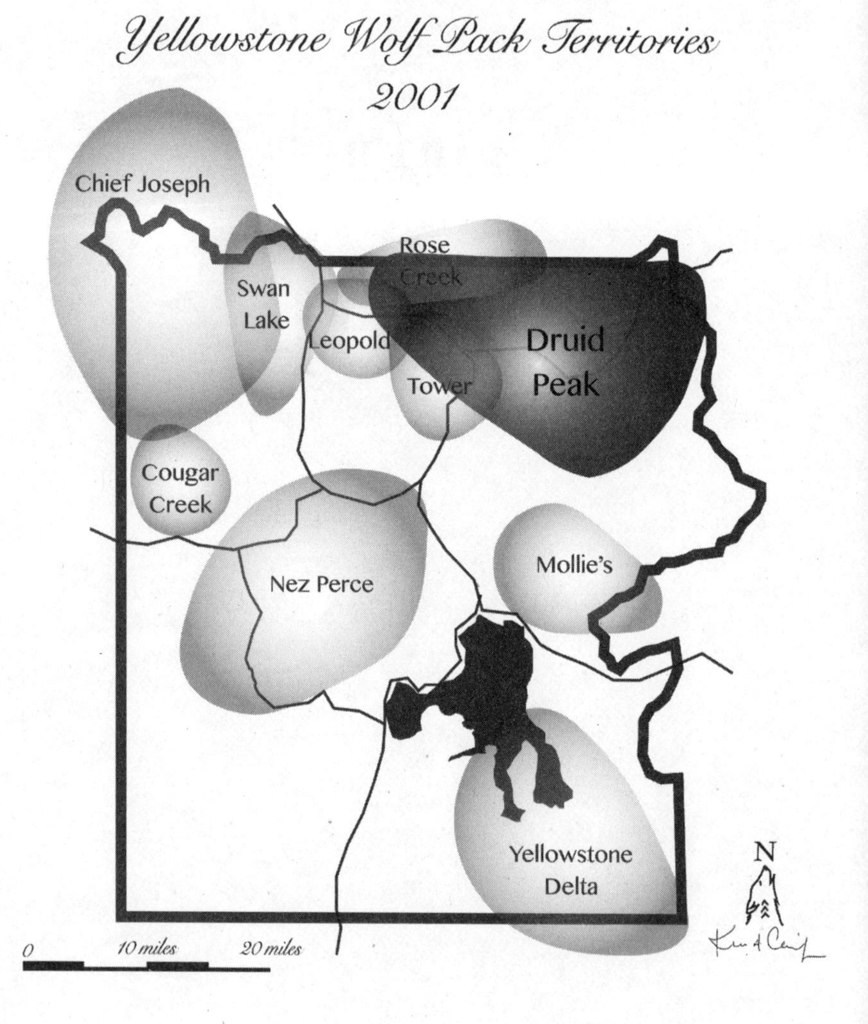
Yellowstone Wolf Pack Territories
2001
Chief Joseph
Swan
Lake
Rose
Creek
Leopold
Druid
Peak
Tower
Cougar
Creek
Nez Perce
Mollie's
Yellowstone
Delta
N
0
10 miles
20 miles

PACKS INCREASE AND decrease in size over the course of a calendar year. These charts show the main pack members in any given year. M=male and F=female. An asterisk (*) indicates a female thought to have denned. The pack of origin for wolves joining from other packs is indicated in parentheses the first time a wolf is introduced. Squares indicate adults and yearlings. Circles indicate pups.

Druid Peak Pack

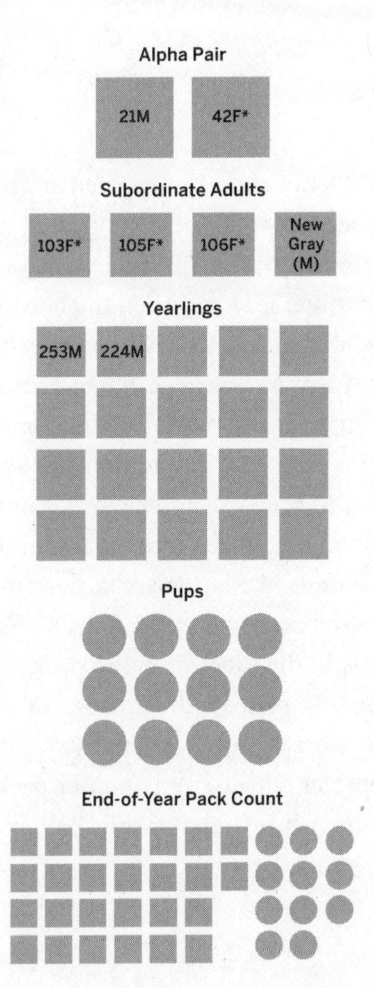

5

It's Complicated

AT THE START of 2001, I anticipated that the approaching mating season would be complicated now that there were two adult males in the Druid pack. During the three previous winters, 21 had been the only male unrelated to the pack's adult females. This winter, New Gray was in the group. I noticed that 21 was beginning to act aggressively to him. I saw 21 run over and pin the younger male.

In a wolf pack with multiple adult males and females, the alpha male tries to breed all adult females unrelated to him and usually attempts to block other males in the group from mating. That sounds like a patriarchal system, but an alpha female might also breed with several males. Years later I saw the alpha female in the Junction Butte pack mate twice with the second-ranking male, right in front of the alpha male. She then took a break from her family, went off on a solo adventure, and met a male from another pack. He bred her twice during their brief relationship. When she returned from her fling, she finally let the alpha male have his turn. That episode turned the old saying "What is good for the goose is good for the gander" into "What is good for the gander is good for the goose."

I thought a lot about why female wolves might behave that way. Within a wolf pack, a mother wolf is dependent on having other adults bring her food and help protect and raise her pups. If two males had bred her, both would likely think that they had sired the pups and would be more dedicated to supporting her than males who had not mated with her. In the case of the Junction Butte female who had a fling with an outsider male, she may have had a reason to think that he had qualities that the males in her pack lacked. She sought him out, then returned to her family, and only then bred with the pack's alpha male. Since he and the second-ranking male had mated with her, both worked overtime to support her and the pups that spring.

On January 1, I saw the Druids farther west than I had ever seen them before. They were near Hellroaring Creek, ten miles west of Slough Creek. The Hellroaring area was part of the Rose Creek territory, and the twenty-seven Druids were pushing farther and farther into the smaller pack's homeland. Now that they were a superpack, the Druids needed a larger territory to support their increased membership and they got that by annexing part of their neighbors' land. No other wolf pack could stand up to them, so the Druids went wherever they wanted when hunting.

The Druids traveled repeatedly to the Hellroaring area. On one trip, seven pups chased a bull elk and nipped at his hind legs and rear end. Naive to the dangers of being right behind an elk, a black pup got kicked and flew backward through the air. On crashing to the ground, the fearless pup jumped up and raced after the elk, eager to have another go at catching him. That kick and the crash apparently caused no significant damage, an indication of how tough and resilient wolf pups are.

By January 10, I saw that 21 was staying close to 42. That day he followed her around and lay down next to her when she bedded. It would still be a few weeks before she would be receptive to mating, but 21 was likely taking no chances that New Gray might try something with her.

In mid-January, the superintendent of Yellowstone, Mike Finley, invited me to a reception at his home at park headquarters for former secretary of the interior Bruce Babbitt, who had approved the park's wolf reintroduction proposal in 1994. I thanked Bruce for that and for making Yellowstone whole again.

The next morning, Mike and Bruce, guided by Doug Smith, would be coming out to see wolves. I got out early that day and acted as a scout for the party, a role I had performed in the past for similar groups. I found six wolves west of Lamar Valley, but they went out of sight before the party arrived. I went back to Lamar and got loud signals from the Druids east of their den forest.

I couldn't spot them and got nervous about the group of dignitaries that would be arriving soon. I hiked through snow up to Dead Puppy Hill and looked for wolves from there. After scanning for ninety minutes, I finally spotted several lumps just barely visible on top of a ridge. One of those lumps raised its head and I confirmed that it was a wolf. I called Doug and told him to bring the party to the

Footbridge parking lot. I rushed downhill and got to the lot just as the cars were arriving. The members of the group took turns looking through my scope and all of them got to see wolves.

A few days after that, Doug did some darting from the helicopter and in two sessions got eight of the twenty Druid pups collared. Four of the adults—21, 42, 105, and 106—still had working collars. The battery on 103's collar had been dead for a while.

In mid-January, Saddleback was away from the pack for two days, then returned to Druid territory. That reminded me of how 21, when he was a young adult in the Rose Creek pack, also left his family for several walkabouts before finally dispersing in the fall of 1997 and joining the Druids. The day Saddleback came back, he showed his value to the pack by killing a cow elk by himself. He had not met up with the other Druids yet. The yearling howled and the rest of the pack howled back. He trotted toward them. A group of pups ran at him and he regurgitated to them. Later the Druids went to his kill and fed.

I saw 42 avert her tail to 21 on January 24 and he sniffed her rear end, a sign that the mating season was beginning. Six days later, he tried to mount her several times, but she was not quite ready, and the mating attempts failed. New Gray came over, probably because he had gotten the scent of 42 coming into season, and 21 chased him away. 21 returned to 42 and the alphas got into a seventeen-minute tie. Eight hours later, they mated again. 21 closely guarded 42 during her days of receptivity and only left her to chase off New Gray. Saddleback was often away from the pack, probably looking for an unrelated female.

On the last day of the month, I saw that 103 was also coming into season. She averted her tail to 21 and he sniffed her, but then walked off. Her scent must have indicated she was not quite ready to breed. Two days later, 21 sniffed 103's sister, 105. She averted her tail and he tried to mount her. By that time, 42 was out of season and likely was pregnant.

New Gray also went to 105 and she averted her tail. Just as he was jumping up into a mounting position, 21 looked over, saw what was happening, and ran back. The younger male immediately got off her and moved away in such a low crouch that his rear end dragged on the ground. He looked like a dog who had been caught doing something wrong. 21 would have to monitor him continually now with three young females in the pack—103, 105, and 106—all coming into breeding condition. 21 repeatedly checked 105 and soon made attempts to mount her. 42 ignored what was going on between them.

21 and 105 mated on February 4. After that he continued to sniff her and keep New Gray away from her. On the sixth, 21 started to sniff 106. He alternated that with checking on 103. Later, when he was totally exhausted, he rested his head on 106's rear end.

We spotted Saddleback bedded down near the pack on February 15. He had last been in the group on January 24. The yearling watched the Druids from a distance and seemed hesitant about approaching them. A black pup ran over to him and tried to interact, but Saddleback ignored it and concentrated on looking at the main group. He howled, and I noticed that he was staring directly at 21, perhaps trying to gauge how he would be received.

Two minutes later, Saddleback got up and moved toward his pack. At that moment, 21 was following 103 around and sniffing her. Then he looked at the young male. That look was enough to turn Saddleback around. He was far enough away that it was probably hard for the other wolves to identify him. 21 and some of the other Druids ran toward him, but soon 21 stopped and the rest of the wolves followed his lead. They must have recognized Saddleback or gotten his scent. The yearling seemed calm and just looked back at them with his tail in a neutral position, rather than tucked in submission. 21 lost interest, went to 106 and sniffed her, then did the same with 103.

I noticed that 42 did not try to block any of the three younger females from breeding, something other alpha females routinely did, probably because they wanted all the meat brought to their den rather than splitting it up among several litters. 42 did not have a domineering personality, and she let the younger, lower-ranking females have more freedom and options than other alpha females might have tolerated. Two of those young females were her nieces and one was her daughter, so that relatedness may have been a factor in how she treated them.

Saddleback continued to watch the pack from a distance. One day three pups went to him. He wagged his tail as they approached, and they all played together. Later he tried going toward the pack again. 106 saw him and trotted in his direction, but New Gray raced past her and charged at the yearling to keep him away from the females. Saddleback ran off, then bedded and continued to watch the pack.

That afternoon, when 21 was bedded down next to 103, New Gray went to 106. She averted her tail and he sniffed her rear end. At that moment, 21 was looking the other way. Then he noticed that the two wolves were together. New Gray tried to look innocent. He and 106 moved apart and 21 lost interest in them. A few minutes later, the young male sneaked back, jumped up on 106, and got in a tie with her. Only then did 21 turn his head and see what was going on. He ran over and pinned New Gray, but it was too late to prevent the mating. 21 stood over them for several minutes, but the two wolves remained tied together, so he gave up and walked off.

That incident showed that a lower-ranking male can breed successfully if he picks the right moment. In any wolf pack that has multiple adults, every male and female wants to breed, and if several of the younger adults are unrelated to each other, it is hard for the alphas to stop them.

103 rushed over and distracted 21. When the tie ended, 21 pinned New Gray, gave him a disciplinary bite, then turned to 103 and attempted to mount her several times without success. It was getting dark by then and I headed in. Although we did not witness 21 breeding 103, she later denned and had pups, so he did get her pregnant at some point.

Saddleback was still in the area and the pups often interacted with him. One day he had a friendly meeting with 42. The yearling joined them at the site and fed with them. They were friendly and playful with him. I last saw him on February 19. I think he realized it was time for him to move on and seek out a female to start a pack with. That brought the Druid membership down to twenty-six. They were still the largest pack in the park since the reintroduction in 1995.

Late in February, the Rose Creek wolves killed an elk west of the Hellroaring area. Due to adults dispersing and low pup survival, the pack was down to just five members. The twenty-six Druids were in the area and must have gotten the scent of the Rose Creek wolves, for they charged in their direction. After getting distracted by some bison, the Druids continued west, this time at a trot. Soon they got the smell of the elk carcass and raced to the site. But rather than feed, the Druids picked up the scent of the Rose wolves and ran west along their trail.

I lost the Druids heading west toward the Leopold territory. I drove to that area and got the Druid signals from Crevice Creek to the north of the park road. That was the farthest west the Druids had ever gone. Their territory went as far east as the Silver Gate area, just outside the Northeast Entrance of the park. The Crevice Creek area was now the other end of that territory, over forty miles to the west. None of the smaller packs in that section of the park could stand up to the Druids. 21 had annexed a huge amount of land, but had done so without using force, just with overwhelming numbers.

There was one group that could be a potential match for the Druids: the Nez Perce pack. They were based on the western side of the park, but occasionally came up to the north. There were twenty-two wolves in Nez Perce at the end of 2000: fifteen adults and seven pups. With the departure of Saddleback, there were now only six adults in Druid, plus the twenty pups. If there was a confrontation between the two groups, the extra adults in Nez Perce could be a decisive factor in which pack would prevail.

6

Separate Dens

ALL THE BREEDING behavior in the Druid pack was over by the end of February, but 21 and 42 continued to have intimate moments together. One day when she was bedded down, he came over to greet her. 42 reached out with a front paw and playfully hit him on the head, then licked his face. When he walked off, she jumped up and did a parallel romp with him, while continuing to lick his face. 21 stopped and she kept licking his face from a pup-like crouched position.

I had seen similar behavior only with the Leopold alphas, wolves 2 and 7, so it was not common to all wolf pairs. Perhaps, as with human couples who have been married for a long time, there is a lot of variation in how affection is expressed, and 21 and 42 happened to have personalities that caused them to act that way. 21 would become legendary as the toughest male wolf Yellowstone ever had, but he was very free in expressing his affection for 42, as well as for his pups.

Nine days later, on March 14, I got 42's signal at the Druid den forest at 6:10 a.m. No other collared wolves were in the area, so she was likely alone there, getting her den ready. If she got pregnant on her first mating with 21, she would be due around April 4. I drove on and found other Druids west of Slough Creek. At 10:55 a.m., I saw 21 and 42 together at Slough. Perhaps she had cleaned out her den, then traveled the nine miles to that area and linked up with 21. I noticed that the sides of her belly were sticking out, a sign of her pregnancy.

I got 42 at the Druid den again on March 18, along with signals from the other collared wolves. The alphas came out of the trees and I saw twenty-one wolves with them. The collar on one pup was giving off a mortality signal, but we had recently gotten a full count of all the pups. Other Wolf Project staff later hiked out to where the signal was coming from and found the collar. It had been chewed off, likely by other pups. That has happened a few other times with Yellowstone collared wolves. In other cases, a mortality signal from a pup might be because the collar slipped off. When researchers put a collar on a pup, they make it a little loose so as the wolf grows up it will not be too tight. If it is too loose, the collar can slip off during play sessions with other pups.

Dave Mech, regarded as the top wolf expert in the world, was in the park that spring. For some time, I had wondered about the type of damage a kick from an elk would do to a wolf. I had watched wolves who had broken legs and seen photos of the broken teeth in the jaws of wolf 8, the wolf who had adopted and raised 21. For a wild wolf, 8 had lived a long life, over six years. His accumulated injuries, incurred during countless fights with elk and bison that were nearly always much bigger than he was, likely slowed his reflexes and contributed to what we think was the cause of his death: drowning in Slough Creek after being kicked in the head by an elk. I asked Dave about how a kick might also damage the internal organs of a wolf. He replied that injuries to organs like the kidneys would probably not be noticed by a biologist examining a dead wolf in the field. A trained pathologist would have to do a full autopsy in a lab to pick up on it and that would rarely be possible.

Dave and I then hiked out to an elk kill. It was a middle-aged cow with a missing incisor and several loose teeth, due to a jaw infection. Often that type of infection is caused by rotting vegetation getting stuck in recesses created by missing teeth. The infection and jaw damage would have made her a vulnerable target for wolves. In a conversation with Dave years earlier, when I was working in Alaska at Denali National Park, he told me that he thought if a full autopsy were performed on every adult prey animal killed by wolves, each one would find evidence of some type of health issue.

In late March, there was a change in the Druid female hierarchy. 106 had always been the lowest ranking of the three adult sisters. I saw her approach 103 with her tail raised high. 106 pinned her, and 103 submissively accepted her sister's dominating behavior. The next day, 106 chased and tackled 103, then let her go.

The following day, 106 chased and pinned 103 once more. 42 came over with her tail raised. That distracted 106, and 103 got up and ran off. But 106 went after her again and tackled her. 42 intervened by walking over and pinning 106. That gave 103 a chance to escape. 106 went after 103 yet again and pulled her down. This time when 42 went to 106, the young female rolled on her back under 42. When 106 finally got up I saw that she was pregnant. I thought that her pregnancy might have made her more aggressive to her sister, as a higher-ranking female tends to get more support from other wolves than lower-ranking ones.

I was intrigued by 42's behavior. I wondered if she was trying to minimize aggression among the females so that they would cooperate better when their pups were born. A few days later, 105 went to 106 with her tail wagging and the two had a neutral interaction. 105 also looked pregnant. Since 105 was the highest ranking of the three sisters, she probably did not see 106 as a threat to her position.

21's personality seemed to pull the pack together and enhance bonding among the family members. On April 4, he was away from the main group of Druids. When he returned, twenty-two wolves rushed over and mobbed him. He responded by greeting them all. It was like watching a person coming home and greeting a group of dogs who were overjoyed to see their human friend.

42 was with the pack in the Hellroaring area early that morning, her possible due date. That was twenty miles west of the traditional Druid den site. She was back at the den in late morning. Her quick traverse of that distance indicated she would be having her pups soon. The next morning, I got signals from 42 and 21 at the site but not from other wolves. 21 must have returned to the den to check on 42. That evening her signal was weak, a sign that she was now underground and probably having her pups.

Four days later, I spotted 21 returning to 42's den from a kill site. Through a break in the trees, I saw 42 pestering him for a feeding. He lowered his head and regurgitated a huge pile of meat. That morning was 105's due date. At first I didn't get her signal anywhere, but that evening I got weak signals from a ridge west of Slough Creek. A tracking flight soon after that reported that 105 was at the den used by Rose Creek wolf 9 in 1996, when her son 21 was still in the pack. The Druids had pushed the Rose wolves out of that area and since a good den site can be hard to find, it made sense for 105 to appropriate it.

I later saw 21 visiting 105. He was likely bringing her food. In the following days, he alternated trips between 42's den and 105's, a round trip of twenty-four miles. 21 was extraordinarily strong and had exceptional stamina, so he could handle it. The other two Druid adult females, 103 and 106, were still traveling with the pack and both looked pregnant. The pups born in 2000 were just over twelve months old now and officially yearlings.

Roger Stradley did a tracking flight on April 18 and called down to tell me that 106 was denning in a forest near Slough Creek, east of 105's site. That location was between the other two dens, so 21 could stop by 106's site on the way to 105's den and when going back to 42's den.

The Druids were at a kill site two days later and both 42 and 105, recovered from having their pups, were in the group. When 105 headed to her den site, 42 followed. I lost sight of them and later got both of their signals in that area. 42 likely was checking to see how 105's pups were doing. I later saw signs that 42, 105, and 106 were all nursing pups.

I stopped seeing 103 with the other Druids on April 21. We eventually found her den in open country west of Slough Creek and downhill from 105's site. The entrance to 103's den was visible from the roadside in Little America, so we potentially could have good views of the pups from there. We now knew that all four Druid females were denning. But in contrast to the spring of 2000, when there were three litters and only six adults, this year there were twenty-two adults to support those four mothers and their pups.

In late April, I was asked to take out a group of Lakota Elders by Rosemary Sucec, the Park Service liaison with Native American tribes who had traditional ties to the Yellowstone area. Rosemary and I brought them to Slough Creek and helped them see wolves, a grizzly, and a black bear. That was the first time most of them had seen a wild wolf. One of the Elders told me how much she admired wolves and how important it was to restore them to the West.

On the first of May, eleven Druids visited 103's den site. A yearling slipped into the burrow, came back out, and shook dirt off its coat. Soon after that, 103 came out of the den and greeted the wolves. They took turns looking in the den and must have been checking out her newborn pups. Two days later, I got signals from the other three females—42, 105, and 106—at the main den in Lamar Valley, and from that time all three seemed to be based there. I couldn't tell if the two younger females had moved their pups to 42's den or if something had happened to their litters. I would have to wait to get a pup count to know more.

I had my first sighting of pups at 42's den on May 6. Through a break in the trees, I got a glimpse of four black pups. One went to 42, who was bedded down, and tried to nurse on her. She jumped up and walked off. The four pups ran after her. We eventually got a count of nine pups there: six blacks and three grays. We still did not know if 105 and 106 had moved their pups to 42's den or if their litters did not survive. Since an average litter in Yellowstone is four to five, there was likely a mixture of at least two and possibly three litters. Later Doug collared four of those pups. Genetic testing showed that one was born to 42 and three to 106. The mothers of the other five pups remained unknown as we never got samples of their DNA.

In trying to figure out why 105 and 106 had moved to the main den, I recalled that elk density in the Slough Creek area was very high in March and early April that year. The Druids spent much of their time there and had a lot of success in hunting elk. 42 left the pack to go back to the den in Lamar Valley on April 4. At that time, elk numbers around that den were low. 103, 105, and 106 all picked den sites where there were a lot of elk, and that seemed to be a smart decision. But that was a temporary situation, for most of the elk soon migrated to Lamar Valley, where they had their calves. 42 had lived in Lamar Valley for five years by then and she had learned how the cow elk concentrated there during the spring. Her den was close to their calving area. To me, that demonstrated intelligence and foresight, along with pattern recognition. The younger females had less experience in elk movements and made the mistake of choosing den sites where elk numbers dropped off later in the spring.

103 stayed put at her den and tried to make do. The younger Druids were not visiting her as often now, probably

because they were busy playing with the pups at the main den. I saw 103's litter for the first time on May 10. She was bedded at the den entrance. Two black pups crawled out of the den, found her, and nursed. After eight minutes, she stood up and ended the session. When she went into the den, one pup followed her. The other pup did not seem to remember where the den was. It crawled around in random directions and ended up downhill from the entrance. The pup must then have sensed the den was uphill, for it started to go that way but the slope was too steep, and it tumbled back downhill. 103 came out of the den and looked around. She sniffed where the pups had nursed, then followed the scent of the missing pup downhill. I saw her reach into the sage and come back up with that pup in her mouth. 103 carried the tiny wolf back into the den.

Four days later, I saw three black pups climb out of the burrow on unsteady legs. Two tripped and fell back into the tunnel. They reappeared and all three tried to walk around the site but often fell over or toppled downhill. The mother bedded down and watched her litter. Eventually all three of them went back into the den on their own. This was her first litter, but 103 was proving to be a good mother, allowing her pups the freedom to explore but still making sure they were safe.

7

The Biggest Pack

I SAW 103'S PUPS playing for the first time on May 21. Two of them wrestled together. Both lost their balance and they tumbled downhill. After that all three pups explored the area near the den entrance. They now could walk around fairly well. That day Doug flew over the area and called down to tell me a big bison carcass was in the water at Slough Creek, just east of 103's den. We didn't know the cause of his death, but hundreds of pounds of meat would be at that site and 103 would surely find it. That carcass could ensure the survival of her pups. I went up on Dave's Hill and saw a grizzly feeding on the bison. 103 would have to be patient and wait for him to leave. She did not have to wait long, for the bear soon departed. 103 went to the site and fed. When finished eating, she went right back to the den. Her pups were still too young to eat meat, but she needed that food to make sure she could produce enough milk for them.

I last saw 103 and her pups at their natal den on June 2. Three days later, a black yearling went to the site and sniffed around. I scanned the general area and saw 21 leading nine Druids to the bison carcass in Slough Creek. After the wolves fed, 42 brought the group to a rocky knoll where they met up with 103. We saw that she had moved her pups to a new den there. Mother wolves often move their pups to new dens and we usually don't know why. In some cases, an infestation of fleas or other insects would be a reason to abandon a site. Other times the mother could have seen a grizzly sniffing around the den area and felt she had to move the pups. The Druids soon traveled back toward the carcass with 103 and 106 trading off the lead position. That reassured me that she chose to den over here, despite the challenges of the location, rather than being forced out of the pack.

I saw fourteen Druids hunting in Lamar around that time. They found a bull elk and chased him. Three yearlings were in the lead and they ran alongside the bull. Another yearling was right behind the elk and got knocked backward by a kick to the head. The older wolves caught up with the lead yearlings and also ran beside the bull. 106 bit into his side. 21 passed her, leapt up, and bit into his throat. He and the other wolves pulled the bull down and finished him off. The wolves that ran beside the elk apparently had learned how that position protected them from being kicked by a hind leg. The wolf that was hit by a hoof hopefully learned the lesson that day.

There was an active beaver lodge upriver from the Druids' main den site. One day the Druids left the den intending to cross the river to reach a recent kill site. 42 was in the lead. When she waded through the water near the lodge, a beaver swam toward her and smacked the surface of the river with its tail to warn its family that a predator was in the area, before diving down and swimming away.

42 and five other wolves waded back and forth in the river, looking for the beaver that made the splash. 21 arrived and joined them in the water. Other wolves stayed on the riverbank and walked around the lodge. One climbed up on the top and looked down the small opening beavers always leave at the top to allow air to circulate. A few wolves went across the river to the carcass and others stayed, trying to figure out how to get the beaver.

At one point, a gray yearling was at the edge of the water, next to the lodge. The beaver swam right at the wolf and it jumped away in surprise. The gray paused, turned around, and saw the beaver staring at it from the water, just a yard away. The wolf backed off. The beaver, now positioned between the wolf and the lodge, swam even closer. When the beaver did another tail slap, the wolf jumped away in surprise, then ran off with its tail tucked between its legs. The beaver swam over to another yearling who was digging at the base of the lodge. The wolf moved toward it. Waiting until the yearling was two feet from the water, the beaver slapped its tail again. That sent water flying through the air. The wolf flinched and ducked its head, trying to avoid getting soaked.

I watched the site for nearly two hours and saw the beaver make thirty-nine alarm splashes as the wolves investigated its home. The wolves never did get that beaver. According to *Wolves on the Hunt*, by Dave Mech, Doug Smith, and

Dan MacNulty, there are no known documented sightings of wolves killing beavers, but researchers have found the carcasses of beavers killed and eaten by wolves. The wolves likely caught them away from water where a wolf would be able to outrun the short-legged rodents. I once saw a great-granddaughter of 21's carrying a freshly killed beaver up to her den. She was coming from a creek and may have snatched it out of a shallow section of water.

In mid-June, I saw a group of Druids at a new bull elk carcass. The yearlings played together and 42 joined in. She chased a black that was running around with a stick in its mouth. Then 21 came over wagging his tail. The yearlings mobbed him. One of them wrestled with him and he pretended the smaller wolf was an equal match for him. After more wrestling, 21 romped off in a playful manner. Later I saw him tuck his tail and run away from another yearling when it charged him, pretending to be afraid, playacting he was not an alpha male. He was now over six years old, about fifty in human years, but still liked to play with his sons and daughters.

One day that week, I got up early as usual and went out looking for wolves, then realized I had done that every day for a full year—365 days in a row.

In late June, we got a mortality signal from the collar on a Druid male yearling to the west of Slough Creek. I waded the creek at a point where the water looked the shallowest, but it still came up to my waist. I found the collar on a ridge farther west. It had slipped off the wolf's neck. When I carried it back across the creek, I thought about the time in late 1999 when I had seen wolf 8 fight an elk in a nearby section of the stream and how he had been kicked and trampled under the water by his opponent. Then I visualized the manner of his death, upstream from where I was, when 8 drowned after being kicked in the head by an elk. This creek had played a major role in the life of the male wolf who had raised 21.

I saw 103 with other Druids in Lamar Valley in late June. That was the first time she had been there since the start of the denning season. Soon after that, I went up on Dead Puppy Hill and saw the nine pups at the main den hunting for voles in the marsh and playing together. With 103's litter, there were twelve pups in the family, along with six adults and twenty yearlings. All that totaled thirty-eight wolves. As far as I knew, that would be the highest number of wolves in one pack ever recorded anywhere. But I had to see them all together for it to be an official record. That meant that 103 would have to bring her pups to the main group.

In early July, I spotted a gray yearling with eight pups in the main den area. It was constantly on the alert as it watched over the pups and followed them around. Usually, when several yearlings were with the pups, some would be playing with them. I wondered if when only one yearling was with the pups, it put off that normal play and supervised them instead. The yearling watched the most distant pups in the group rather than the closer ones, showing that she was most concerned about the ones farthest away. When all the pups were closer together, the yearling constantly looked from one pup to the next one and on to all the others, checking each in turn.

The yearlings took their role as babysitters seriously and they were good at it. I later saw two yearlings take charge when a grizzly wandered into the marsh below the den forest. Five of the pups were there, along with a black yearling. As the pups played and ran around, the black approached the bear. A gray yearling saw the situation, came down to the marsh, and headed toward the grizzly as well. The pups were playing nearby and seemed unconcerned. The yearlings and grizzly moved slightly out of my sight. Then I heard the bear growling. The pups immediately raced toward the den and probably hid inside. The grizzly ran back into view, with both yearlings chasing it away. It seemed like the pups knew instinctively to run underground when they heard that growl.

The Druids regularly made trips to Slough Creek to visit 103 and her three pups. One wolf watcher told me she saw 42 arrive when 103 was not there. The three pups came out and 42 regurgitated to them. Just as one wolf mother will nurse another's pups, she will also regurgitate to them when they are old enough to eat meat. 103 might have chosen to den away from the main site, but 42 was going out of her way to help the lower-ranking female.

A black pup at the main den followed some of the adults across the road to the Chalcedony Creek rendezvous site on July 19, and the next day two gray pups joined that pup. The nine pups were hard to keep track of after that. Doug did a flight on the twenty-seventh and saw seven pups and eleven adults up high at the Opal Creek rendezvous site. He got the signals from the alphas farther to the south, so the younger adults must have been taking care of the pups. We didn't see any pups at the main den after that, so they must all have been moved to one of the rendezvous sites.

I went to Slough Creek nearly every day to check on 103's three pups. 103 regularly made round trips from her den to Lamar Valley. On August 1, I saw her arrive at the Chalcedony rendezvous site and regurgitate to two pups there, pups that might have born to 42. Earlier 42 had fed her pups and now she was returning the favor by feeding pups born to another mother.

The next day, the rendezvous site at Chalcedony Creek was full of wolves. I eventually counted twenty-five: sixteen adults and all nine pups from the main den. The pack was especially playful that day. 21 reared up and put his front paws around 42's shoulders. A gray yearling caught a vole, let it go, chased it, then pushed it around with her nose. Like previous generations of Druid pups, the new ones found the marsh, where voles were plentiful, and spent a lot of time catching and playing with them.

Early every morning in August, I checked for wolves in Lamar Valley, watched if they were visible, then went to Slough Creek to look for 103 and her pups. She was often away from her litter, hunting for food to bring to them. Other Druids visited regularly. I noticed that New Gray seemed to be with 103's pups more than any other Druid wolf. The pups always ran and greeted him like a favorite uncle, and he always regurgitated to them.

A bison died of natural causes in Lamar Valley and the Druids found it. On August 11, I saw ten of the adults there, including the alphas. A big grizzly had taken over the carcass and was keeping the wolves away from it. 21 was standing off to the side, next to an object that seemed to be part of the bison. Later he went over to the bear, growled at it, and nimbly dodged when the grizzly charged and swung a front paw at him. Then he worked out a compromise with the bear. 21 and six other Druids fed on the opposite end of the carcass from the grizzly. This was probably an old bear who would rather concentrate on feeding than waste a lot of time repeatedly chasing off wolves who, it knew from experience, would just come right back. Perhaps this grizzly and 21 had known each other for years and had made these deals many times before.

21 later went back to that separate object and plucked fur from whatever it was. I could tell from its size and shape that it wasn't from the bison carcass. When 21 lifted it up, I saw that it was the remains of a first-year grizzly cub. I called Kerry Gunther, Yellowstone's bear biologist, and he arranged to send two of his staff out to Lamar to investigate.

I looked at the adult grizzly again. People on the scene told me it was a female. I noticed two places on her coat where fur was missing, likely injuries from a fight with another bear. Later that day, I met up with bear biologists Travis Wyman and Susan Chin. They were carrying the grizzly cub. It was a forty-pound female and she had puncture wounds on the throat. Several of her ribs were broken. That must have been caused by a blow from an adult bear, probably before it bit the cub's throat. The fur and skin were gone from her abdominal area, but nothing seemed to have been consumed there or elsewhere.

I interviewed Bill Hamblin, a regular Yellowstone visitor from Idaho who was an expert on grizzlies, and he told me he was first on the scene early that morning. In addition to the female bear at the carcass, a second grizzly was leaving the area. He had probably killed her cub. The Druids came along after that, and 21 found the dead cub but did not feed on it. It is not unusual for male grizzlies to kill cubs. If a mother grizzly loses her new cubs, she may go right into breeding mode so she can have a replacement litter. The male bear that kills her cubs could be the one that mates with her. Therefore, he could replace another male's cubs with his own. That behavior contrasts with male wolves who will adopt and raise pups born to another male who has died, as 8 did when he adopted 21 and his seven siblings, and as 21 did when he raised the five Druid pups after joining their pack.

In mid-August, I saw all twenty-six adult Druids in Lamar Valley, mostly at the Chalcedony rendezvous site. Five pups were also there. That added up to thirty-one wolves. I went to Slough and saw one of 103's three pups. I was still trying to get a count of all twenty-six adults and twelve pups together. I heard that pilot Roger Stradley had seen all three of the pups at Slough recently, so they were doing well.

A few days after that, when two black pups were together in Lamar Valley, I noticed that one was much bigger than the other. The markings on the smaller pup matched those on one of 103's pups. We had estimated that her pups had been born about eighteen days after 42's, so that would explain the difference in size. That sighting established that at least one of 103's pups was now in the main group in Lamar. 103 was in the group that day, as well, and the small pup was bedded down close to her.

On August 29, I finally saw what I was looking for. That morning I counted twenty-three wolves at the Chalcedony rendezvous site. Then another fourteen came in and joined them. That added up to thirty-seven wolves in the group, the world's record for any known wolf pack. Since all twelve pups had survived, the count should be thirty-eight. I later went to Slough and saw a black pup there, the thirty-eighth Druid wolf. I never had that many of the Druids together after that date. Some of the adults and yearlings soon began to disperse and attempted to form new packs. Eventually that last black pup at Slough joined the other eleven at the Chalcedony rendezvous site. That meant that as of early September, all twelve known pups had survived.

The pups from the different litters were getting along well together. One day I saw a small black pup, likely born to 103, chasing a much bigger black pup and nipping it on the rear end. The big pup tucked its tail between its legs, then chased the little pup. They reversed roles once more and the smaller one chased the bigger one, nipping its rear end and hind legs.

Waterbirds were often in a marsh near the rendezvous site, and the pups were fascinated by them. One day four pups approached forty Canada geese. Most took off, but one goose stayed on the ground and the lead pup ran at it. When the bird flew off, the pup sniffed where it had been. On another day, a black pup repeatedly approached a northern harrier, formerly known as a marsh hawk. The pup walked toward the hawk wagging his tail, signaling he was hoping to play with the bird. As he got closer, the hawk flew off a short distance and landed. The pup eventually approached six times and the harrier always flew away. On the last approach, the pup leaped up and tried to grab the bird as it flew over him.

September 11, 2001, started off just like any other morning. I happened to have my car radio on and heard the news of the attack on the Twin Towers in New York City. Doug was doing a tracking flight with Roger and they turned back when they heard that all planes across the country had to land immediately. Years later a Yellowstone male wolf was collared and assigned the number 911. When he was an old wolf, 911 died in an extraordinarily heroic manner, an act that reminded me of the courageous firefighters, police officers, medical personnel, and regular people who tried to help victims of the attack and lost their lives in the process. I will tell 911's story in another book.

In early October, I saw 42 and 103 chasing 105. Nearby, 21 and other wolves were chasing New Gray. 21 nipped him on the rear end a few times during the pursuit. A few days later, 105 and New Gray were gone. 105 came back in late October without New Gray. 42 led some of the other females in driving her away. After that 105 did not return to the pack. We never saw New Gray again. Around that time, I saw 42 pinning both 103 and 106. They, along with 105, were four and a half years old, about thirty-eight in human terms. Most young females would have dispersed from their families well before that age. It was time for the three sisters to find mates and start their own packs.

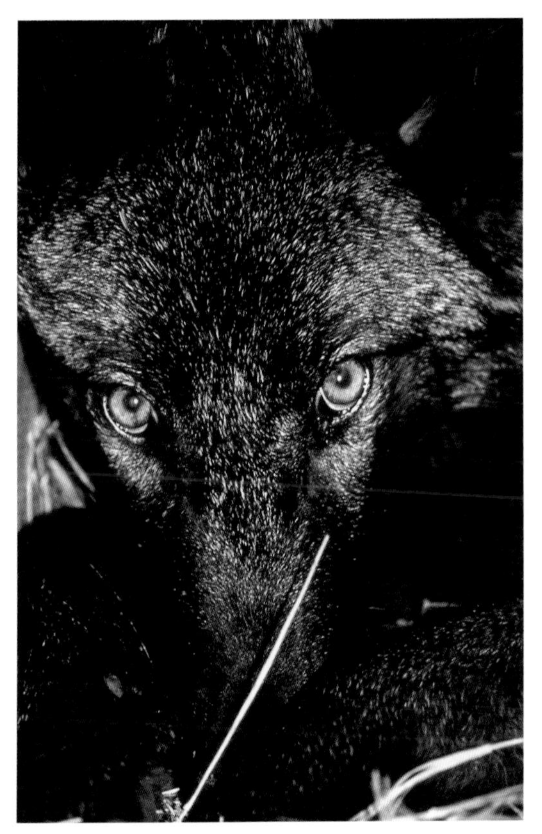

21 at five months. His pack was relocated to an acclimation pen after the alpha male, 21's father, was shot. **NPS/Barry O'Neill**

21 at eight years old with a lot of gray on his black coat. He had been the alpha male of the Druid pack for over five years. **Doug Dance**

The Druids chasing bull elk. Wolves are pursuit predators and prefer to make a running attack on their prey. **Doug Dance**

Leopold wolves hesitant to approach elk that are standing their ground. **NPS/Matt Metz**

21 investigating a coyote den and reacting when one of the adults pulls his tail. Wolves have a contentious relationship with coyotes, for they often steal meat from wolf kills. **Elva Paulson**

Wolf 42, longtime mate to 21, when she was eight years old. Her once-black coat had turned gray. **Diane Hargreaves**

21 looking back at 42 and the rest of the Druids. Wolf 253, a valued pack member despite having a permanent leg injury, is fourth in line. **Diane Hargreaves**

A Druid pup scratching its back by rolling on the snow. With their large paws and thick coats, wolves are superbly adapted to winter weather. **Bob Weselmann**

21 confronting a grizzly who had taken over an elk 21 had killed. 21 would sneak up, bite a bear on the rear end, and get it to chase him. 21's family then would run in and feed. **Betsy Downey**

The Slough Creek wolves howling. The pack was cofounded in 2002 by wolf 217, one of 21's daughters. Her mate was from Mollie's pack.
Diane Hargreaves

The Agate Creek alphas. The upper wolf is 472, another of 21's daughters. She and 113 from the Chief Joseph pack founded the Agates, also in 2002.
Bob Weselmann

Wolf 302 on his first day near the Druid wolves in early 2003. He had a personality totally different from that of 21, his uncle. The young Druid females found him irresistible. **Doug Dance**

302 with one of 21's daughters that same day. 21 had to raise five pups born to 302 and his daughters that spring, along with pups 21 had sired. **Doug Dance**

Photos of 21 and 42 on their last day together. They were nearly nine years old, twice the average life-span of Yellowstone wolves, and had been a pair for 70 percent of their lives. **Photographs by Kim Kaiser**

8

The Battle of Lamar Valley

ON OCTOBER 19, the Nez Perce wolves, the group based on the western side of the park, invaded Lamar Valley. There were eighteen adults in that pack, fewer than the twenty-six in the Druid pack, but the Druid adults were not always together. Every wolf in the Nez Perce pack was gray, while the Druids were mostly blacks.

In early morning, I saw 21 with six other Druids bedded down at the Chalcedony Creek rendezvous site. I got signals from 42 and spotted her and ten other Druids two miles to the west. As 42 and the wolves with her traveled toward 21's group, a cluster of ten Nez Perce wolves exploded out of the trees. They collided with 42's group and there was a confusing melee of wolves running back and forth.

42 stepped away from the conflict and howled. The Druid wolves at the rendezvous site howled back. The message

42's howls conveyed to 21 must have been a call for immediate help, for his band of wolves were already up on their feet and traveling west at a fast pace. 21 was in the lead. He jumped up on his hind legs to get a better look at what was ahead.

42 and the wolves with her ran toward 21. He was clearly in charge of defending his family and ran out in front of his group. Then I heard constant howling from the trees to the southwest, indicating that the Nez Perce wolves had retreated into the forest. They must have seen 21 running in with reinforcements and fled.

I put my scope on 42 and saw blood on her hip. That must have come from a bite inflicted during the early part of the interaction. 21's group joined 42's. 21 would have immediately noticed that wound, sniffed it, and gotten the scent of rival wolves. Knowing him, I figured that made him determined to drive the Nez Perce out of his territory, and so the Druids kept on running forward. They soon stopped and had a group howl. After failing to find any lingering enemy wolves, they turned around and headed back toward their Chalcedony rendezvous site.

At that point, assuming the confrontation was over, I went west to look for the main Nez Perce group, which meant that I missed what happened next. Fortunately, Cliff Brown, a longtime Cooke City resident, filmed the incident and later lent me his tape.

The early part of the footage showed eight Nez Perce wolves, who must have just come out of the trees, chasing several young Druids. The Nez Perce wolves mainly stuck together while the Druids scattered. After a few minutes, the

Nez Perce stopped and had a loud group howl. They intently looked east, then charged in that direction.

Cliff's footage next showed a group of sixteen Druids running directly at the Nez Perce wolves, with 21 out in front. He was aiming straight at the tight cluster of rival wolves, who had stopped. It looked like he was planning to crash right into them. I got the impression that 21 did not care if any other wolves were following him. His family was being threatened and he had no problem with charging alone into the wolves who had wounded 42.

A moment later, when 21 ran into them, the Nez Perce wolves fled without putting up a fight. They were too intimidated to engage 21. Cliff's video then showed the Druids chasing the Nez Perce wolves back and forth. Totally disorganized and scared, the intruders ran off into the trees to the southwest.

The rest of the footage showed 21 and 42 sniffing around the area where the Nez Perce wolves had been. Then 21 led his group into the trees where we had lost the invading wolves. The Druids failed to find any rival wolves and reappeared. The pack ran a patrol around the area. At times they looked west and howled, then resumed their search. It looked like they were conducting a military mopping-up operation, making sure no enemy troops were hiding in the area.

I will never forget that moment in Cliff's video when 21, out in front of the other Druids, ran into the cluster of eight Nez Perce wolves. It was a stunning demonstration of bravery in the middle of a battle. That image of a father wolf risking his life to protect his mate and family forever defined the character of 21 for me.

The following morning, the Druids were back at the Chalcedony rendezvous site. A black yearling got up and I saw that his left hind leg was injured. He had likely been bitten during the initial confrontation with the Nez Perce. Later he walked around a bit but put no weight on his bad leg. 103 and a gray pup came over and had a friendly greeting with him. The injured yearling wagged his tail at them. I saw that he was having a hard time with any movement, either walking or trying to lie down. The other wolves left, but later in the day, 21 came in and brought food to the yearling. The young male stayed bedded down at the rendezvous site during the coming days, mostly alone, not doing much other than occasionally lifting his head and looking around.

The pack returned on October 24 and I got a count of twenty-nine wolves at the rendezvous site. Most had full bellies and I saw one wolf give a piece of meat to a pup. I hoped that the injured yearling got a share of the food. I later saw him walking around with other wolves. He was still keeping that hind leg off the ground.

Two days later, I found some of the young Druids returning to the rendezvous site from a fresh elk carcass in Lamar Valley. The lead wolf, a black pup, was carrying a leg from the carcass. When it arrived at the rendezvous site, that pup dropped the leg next to the injured yearling, wagged his tail, then greeted him submissively. The pup moved off, leaving the leg for its packmate. That pup was used to adult wolves giving it food and now it had matured to the point where it was giving food to its injured older brother. I was really touched by its behavior.

A few hours later, the alphas came back to the rendezvous site and all the other wolves ran to greet them. The injured black somehow kept up, running on just three legs. The next day, October 27, eight days after the battle with the Nez Perce pack, that wolf left the rendezvous site with his family when they traveled west and kept up with them, holding his wounded leg off the ground. He often stopped for a brief rest, then always continued on. That black was later collared and became known as wolf 253. Despite a permanent disability, he lived a life of great accomplishments. Later, when his DNA was analyzed, it showed that he had been born to 105 and was sired by 21. He would have been one of the pups the females saved from 40 in the spring of 2000.

In November the Druids often traveled from Lamar Valley to the Hellroaring area and back, a round trip of forty miles. One day they were bedded down about midway between those locations, in Little America. There were twenty-one Druids, including a lot of pups. 21 found an elk leg and chewed on it. A black pup came over and, wanting to play, hit his father on the head with a paw. 21 did an air snap at him, which the pup ignored. It rolled on the ground in front of the big wolf and hit him in the face with both front paws. At that point, a raven walked over and 21 jumped up to chase it off. The pup grabbed the elk leg and ran off with it. 21 good-naturedly let the little guy keep his stolen prize. Soon after that, I noticed that 253 was playing with a pup, a sign he was feeling better.

Later that month, there were twenty-five Druids in the Hellroaring area, including the alphas. The wolves went west with 253 leading, still using just three legs. 21 took the lead position, but soon 253 passed him and led once more. I had come to really admire 253 for his refusal to allow his disability to hinder him.

253 reminded me of a story that Doug Smith once told me. He was in the helicopter, trying to radio-collar an alpha female who was running out in front of her pack. Taking careful aim, he got a dart in her. After the helicopter landed, Doug walked over to the semiconscious wolf and began examining her. Everything seemed normal until he looked down at her legs. He was startled to see that there were only three. Yet she had run from the helicopter faster than any other wolf in her pack. The wolf had probably stepped into a wire snare outside the park and either escaped with the wire still wrapped around her leg so tightly that it cut off the blood supply, or chewed off her leg, something that has been documented in other wolves. Thinking about that makes you realize what a wolf is willing to do to be free. The motto of New Hampshire is Live Free or Die. It's also the code wolves live by.

The Druids were approaching a hiking trail at Hellroaring that Doug and a co-worker had traveled along just thirty minutes earlier. 21 stopped when he intersected the trail and seemed especially interested in sniffing at it. There was one spot that he smelled three times, like he was trying to place the scent. Doug was the person on the rescue team who pulled the eight Rose Creek pups out of 9's den back in the spring of 1995. I think that dogs always remember the scent of dogs and significant humans they have met, and that made me wonder if 21 recognized the scent of the man who had rescued him.

Doug had been with the Yellowstone wolf reintroduction project from the very beginning in 1994 and took over the leadership position in 1997. He was an experienced wolf biologist who was passionate about his subject. I always tried to attend talks Doug gave on wolves because he could elegantly explain complicated wolf research to regular people and do it with enthusiasm, better than I could. He was especially skilled in pulling all the elements of his talk together at its conclusion in a way that inspired and motivated his audience to care about wolves. I once heard him say, "I cannot wait to tell people cool things about wolves."

For many years, I was the only person on Doug's staff who didn't know how to text. He jokingly told me I was a dinosaur. I thought about that. Dinosaurs never learned to text and look what happened to them. Hoping to avoid their fate, I finally figured out how to do it. Then I stopped looking up in the sky for asteroids heading my way.

Every spring Doug would get a memo informing him that all his staff had to go through mandatory security training for using government computers. Doug had to tell the bureaucrat that he had one guy who didn't have a government computer and never used one. He had to go through that sequence year after year. Doug probably didn't mention it, but I also didn't have a government office, a government desk, or a government phone. My office was Lamar Valley.

Around that time, I came across a book that helped me get a better understanding of how wolves might think. In *Thinking in Pictures and Other Reports from My Life with Autism*, Temple Grandin explains that she thinks in pictures rather than in words. She is a professor of animal science at Colorado State University and does consulting work designing animal-holding facilities. When she is asked to plan a new facility, she sorts through a series of mental images of other facilities she has seen or worked on and then creates a slideshow in her mind of the features that would work best for her new project.

That explanation of how Temple's mind worked made me realize that wolves must think in pictures as well. I recalled a time when I was watching wolf pups at Slough Creek. One pup had a piece of meat and ran off with it. She zigzagged around the rendezvous site for some distance, found a spot out of sight of the other pups, buried the meat there, then romped off. Another pup had watched her carry off the food and must have suspected that she was going to hide it. That pup got up, found the scent trail of the female, even though it was mixed in among the scent trails of many other pups, followed her circuitous route to the hiding spot, and dug up the meat. I imagined that when he first came across her scent trail and sniffed it, he got an image of that specific pup in his mind, matched it with the appearance of the pup with the meat, and followed her trail. If it crossed the trail of other pups, he would have sniffed at those scents, gotten visualizations of other pups, then made sure he went back to the female's trail.

The same process would work for older wolves. If 21 returned to 42's den with meat from a new carcass and gave it to her and the pups, she could later sniff around the nearby scent trails, find one that triggered an image of 21 in her mind, then backtrack that trail. If that route crossed the trail of other wolves, she could sniff at each one and get images of each packmate that made those trails until she found the one that created a picture of 21 in her head, and then follow it.

On December 8, the Druids were well west of the Hellroaring area. I got signals from the Rose Creek wolves farther to the west. The much larger Druid pack was pushing the Rose wolves out of their original territory. Doug was flying that day and he called down to tell me he counted twenty-three wolves in the Druid group and nine in the Rose pack. The Leopold wolves were in their territory to the southwest and he got a count of fifteen.

Based on Doug's information, I hiked to a viewpoint where I could see both the Druids, who had bedded down, and the Rose group farther to the west. I soon heard faint howls from the Rose wolves. The resting Druids responded by jumping up and having a loud group howl. At the sound of the larger pack howling, the Rose wolves ran off. Tom Zieber was on the Leopold winter study crew and he radioed me to say his pack must have heard the Druids as well, for they howled back for ten minutes from their territory.

At that time, there would have a been a male yearling in the Leopold pack born to 21's older half sister, wolf 7. She left the Rose Creek pack soon after her family was released into the park in 1995. She met up with wolf 2 from the Crystal Creek pack and together they started the Leopold pack, the first new pack to be formed in Yellowstone after the wolf reintroduction. Her mother, wolf 9, gave birth to 21 after she had left.

The group howl by the Druids would have been the first time 21's nephew became aware of the existence of 21 and his pack. When older, that wolf would be collared and given the number 302. He would travel to Lamar Valley and become well known to his uncle, but not in a good way.

The following day, while 21 and 42 were resting, 106 and some of the yearlings and pups chased an elk herd. They singled out a cow and killed her. The alphas and other adults joined them, and the younger wolves shared their kill with them. That was becoming a common occurrence in the large pack. Young wolves, eager to be active, often made kills without any help from the older wolves. All the hard work 21 had invested in raising and training those wolves was now paying off and making his life easier.

The most valuable player in that hunt was black yearling 224. He got a holding bite on the cow's hind leg and held on, despite her attempts to kick back at him. Then three other wolves ran in and helped him finish her off. But later, despite his service to the pack, other yearlings ganged up on 224 and picked on him. After those wolves walked away, 21 went to the bedded 224 and the yearling licked his face. 21 licked him in turn, a reciprocal gesture that I could not recall 21 giving to another male, then went to the carcass. 224 followed his father. No wolves bothered him.

Because of his size and fighting ability, 21 probably had never been picked on, but I think he had seen what happened to 224 and had empathy for him. That incident, and the one where he came back to check on the injured 253 and brought him food, showed how 21 watched over the younger wolves in the pack and gave special attention to ones that were having a hard time, like a human father would to a son or daughter.

PART III

2002

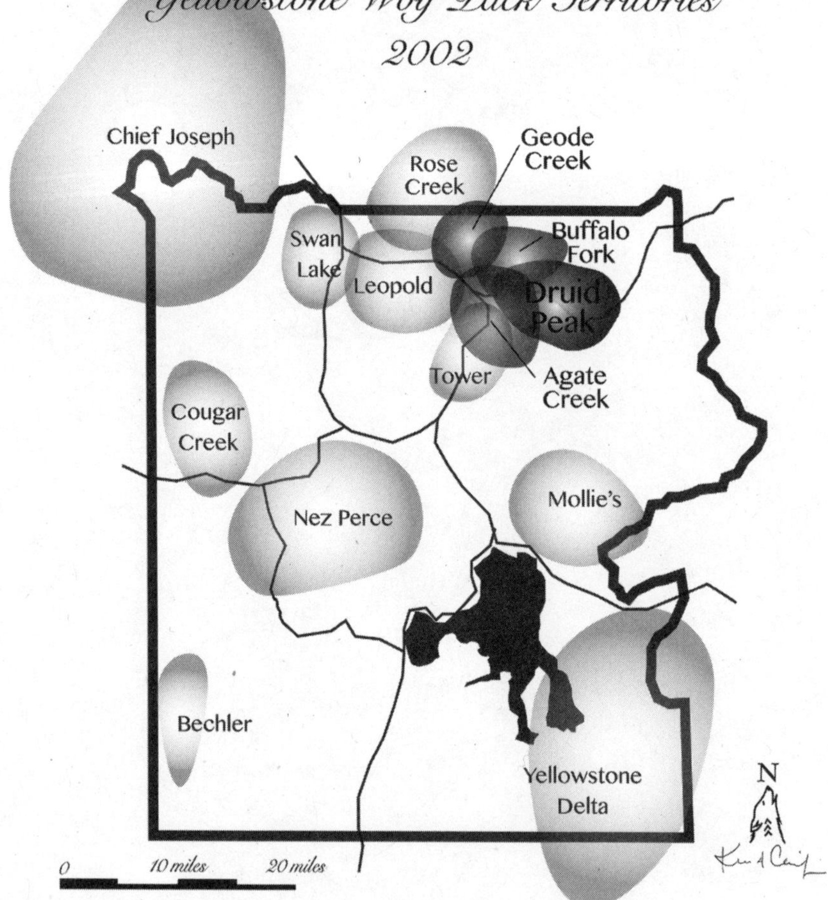
Yellowstone Wolf Pack Territories
2002
Chief Joseph
Rose Creek
Geode Creek
Swan Lake
Leopold
Buffalo Fork
Druid Peak
Tower
Agate Creek
Cougar Creek
Nez Perce
Mollie's
Bechler
Yellowstone Delta
N
0
10 miles
20 miles

PACKS INCREASE AND decrease in size over the course of a calendar year. These charts show the main pack members in any given year. M=male and F=female. An asterisk (*) indicates a female thought to have denned. The pack of origin for wolves joining from other packs is indicated in parentheses the first time a wolf is introduced. Squares indicate adults and yearlings. Circles indicate pups.

In 2002, several young Druid females permanently joined new packs. Other Druid females were temporarily in those packs, then returned home.

Druid Peak Pack

U Black, originally a Druid, was temporarily a member of the Agate Creek pack, where she became pregnant. She later returned to the Druids and had her pups in Lamar Valley.

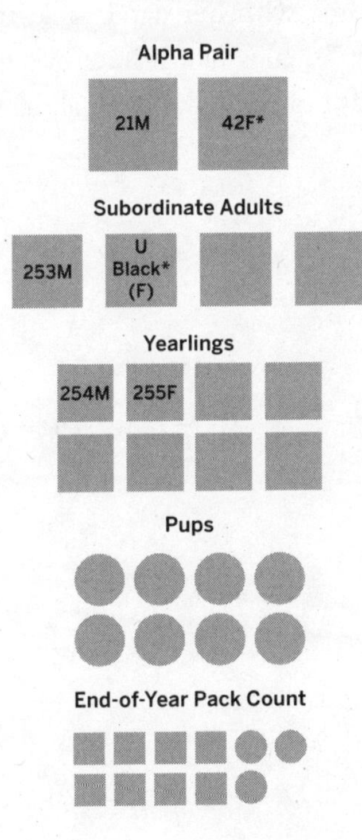

Buffalo Fork Pack, Formed 2002

218 spent time with both the Agates and the Geode Creek pack before finally settling down with the Buffalo Fork pack.

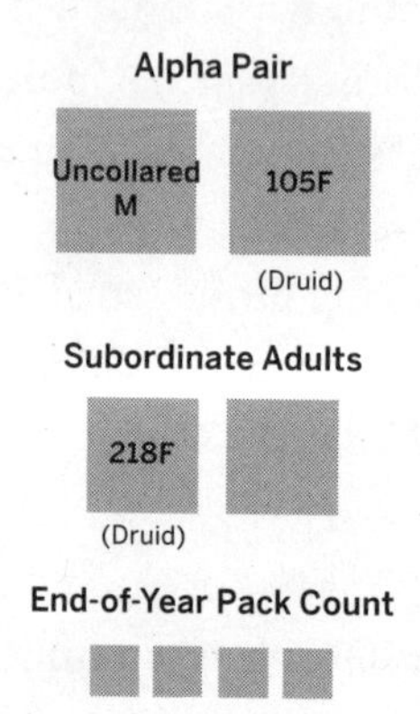

Agate Creek Pack, Formed 2002

251 often spent time with her natal pack, the Druids. Two Agate Creek pack females had litters; one was 103 and the other was either 251 or the uncollared alpha female.

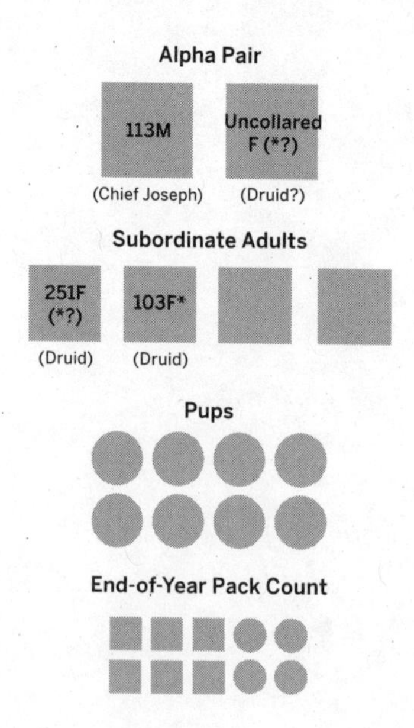

Geode Creek Pack, Formed 2002

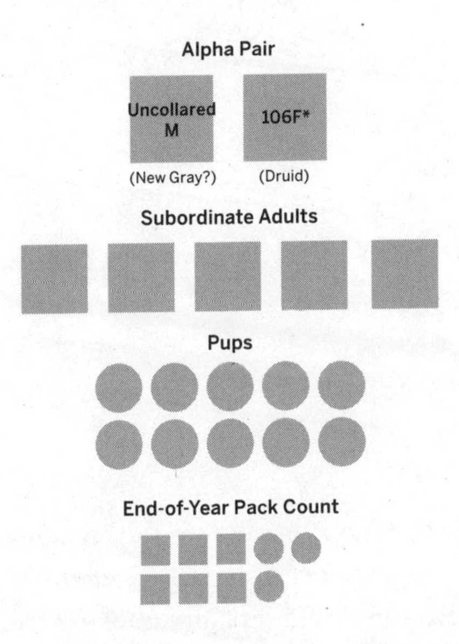

Slough Creek Pack, Formed at the End of December 2002

217 spent time with the Agates and the Geodes and returned briefly to the Druids before finally settling down to start this pack at the end of the year.

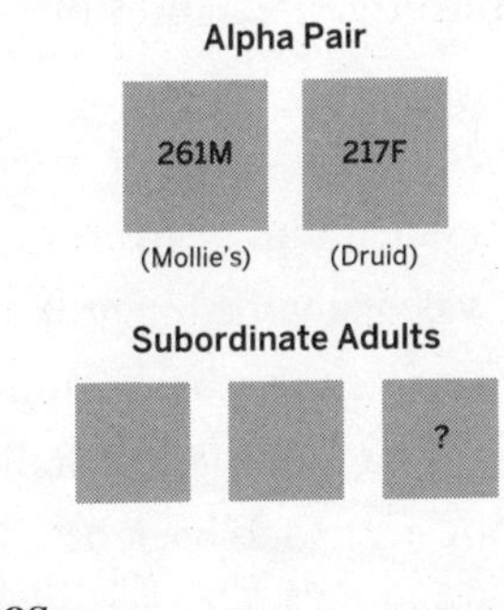

Unattached Males

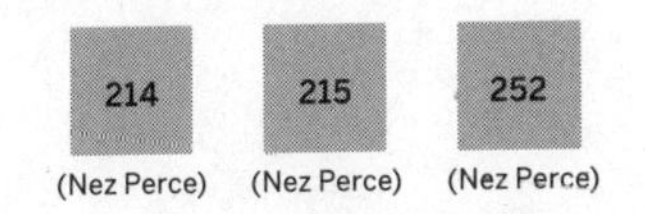

9

The New Packs

TOWARD THE END of 2001, as the breeding season was approaching, we began to see males from other packs come into the Druid territory, seeking females to pair off with. On the day before Christmas, I found a big collared gray in Lamar Valley. It was 113 from the Chief Joseph pack, forty miles to the west. I had watched his family for a month in August 1998 when he was a yearling. Now he was with one of the black Druid females. When 21 and seventeen other Druids came into the area and howled, 113 discreetly slipped away.

When he came back into sight, he met up with a male who had a dark gray coat. Judging from the dark gray's submissive behavior, he was likely the big gray's younger brother. The black Druid female went toward both males wagging her tail. 21 noticed what was happening and stalked that way. The two males ran off. I put my scope back on 21 and saw that his hackles were up, making him look extremely intimidating.

The big gray Chief Joseph male, 113, soon linked up with the black Druid female again. She wagged her tail, sniffed his fur, and did jump feints and play bows to him. After that she jumped on his back. The young female was acting just as 42 had done when she first met 21 years earlier. Another black female ran to the big gray and was just as welcoming. Before things could progress any further, 253, the Druid male who had been injured in the confrontation with the Nez Perce pack, came on the scene. 113 ran away, but the two sisters followed and when he stopped, they continued flirting with him. Soon two more black females joined the group, and both seemed equally interested in this new male. 21 and 253 went toward the gray male with their hackles raised. The newcomer ran off, but one of the four females soon joined him and continued to flirt.

By that time, 21 had given up trying to control his yearling daughters. He went back to 42 and bedded down next to her. Those Druid females were about seventeen in human years. 21 had a lot of daughters that age, perhaps as many as ten. He was like a king in a Disney animated movie with ten beautiful daughters, trying to cope with suitors flocking into the kingdom from far and wide.

On the first day of 2002, I saw Chief Joseph male 113 and his dark gray brother with two black Druid females. 113 was clearly the alpha of the group. He did a raised-leg urination and the dark gray male did one too, which indicated he was likely older than a yearling. One of the black females also did a flexed-leg urination, a sign she was the alpha female of the group. We occasionally saw a third black female with the original four wolves. It took some time for the membership to settle out, but all the females seemed to be Druids. We eventually called the group the Agate Creek pack. Their territory included the west end of Specimen Ridge and the area around Tower Junction.

Other wolves had tried for years to establish territories there, but they had all failed, due to the proximity of larger, well-established packs that treated newly formed groups aggressively. 113 had a leadership style like 42's, meaning he was skilled in getting the adults in his group to work together. He never seemed threatened by having other males in his pack, including ones that appeared unrelated to him, and he treated them well. In turn they respected his alpha male status and helped him contend with neighboring packs. The Agates would be a major pack for many years to come, primarily thanks to 113's social skills.

When I watched 113's original family, the Chief Joseph pack, in 1998, it consisted of the alpha pair, four gray male yearlings, and seven pups. One of those males was especially interactive with the pups. In my field notes, I referred to him as the playful yearling and felt that he had the makings of an alpha male. As I studied how 113 formed and led the Agate pack, I thought that he could be that yearling. When he later had pups, I saw that 113 behaved with them the way that yearling had interacted with the Chief Joseph pups. The similarities in behavior made me think they were likely the same wolf.

Meanwhile, Druid female 106 was also striking out on her own. A few days later, I saw her with four grays in the Hellroaring area, west of Lamar Valley. Her group became known as the Geode Creek pack.

For a while, Druid yearling females 217 and 218 alternated their time between the Agates and the Geodes. 218 finally settled down with her sister, 105, when 105 founded the Buffalo Fork pack that spring with an unknown alpha male. 217 eventually returned to the Druids, but later started a pack of her own. That fluidity in newly forming packs turned out to be common. Young wolves leave their families with siblings and may be in one new group for a while, then try associating with relatives in other newly forming groups. Perhaps in the end the main issue is finding a group of wolves that they get along with. It would be like a young guitar player trying to find the right group to start a band with.

In January my counts of wolves in the main Druid pack tended to be twelve to fourteen. The alphas and 253 were always in the group, along with a lot of pups. That meant that most of the yearlings had dispersed. I saw 42 avert her tail to 21 for the first time that season on January 27. He licked her face affectionately. The two mated on the last day of the month. One day I saw the pack bedded down and noticed a big lump near them. It lifted its head and I realized that it was 21. He was so large I had mistaken him for a bison.

The Agates gradually acquired more members. 103, now middle-aged, left the Druids to join her younger sister, who was the pack's alpha female, and 113 bred 103. 214 and 215, two brothers from the Nez Perce pack that had got into a confrontation with the Druids in Lamar the previous October, were in the area looking for Druid females. They found some in the newly forming Agate pack and 113 allowed them to join the group. On February 9, I saw 214 mating with an uncollared black female with a parenthesis-like area of gray fur at the back of both hips. After 214 mated, he and his brother wandered off.

In the middle of February, as I was driving through Lamar Valley in the early morning, I saw two gray Druid pups on the road, walking toward me. I backed up to get out of their way, but they ran after my car. I stopped, got out of my vehicle, and yelled and clapped my hands, hoping that would scare them into leaving, but one pup continued toward me. I made a snowball and threw it toward the pup. It walked over and sniffed it. That made me think tourists had recently thrown food at those pups. If so, it would be a bad development. Some wolf watchers had stopped their car near mine and I asked them to sound their horn. Finally, the pups left the road and walked off.

Brian Chan was the law enforcement district ranger in Lamar Valley, and I reported the incident to him. He told me that a month earlier he had had a similar incident with a black Druid pup on the road. He had fired two slingshot pellets at the wolf and each time the pup walked over and sniffed the pellet. That was another sign people had tossed food at the pups. I called Doug Smith and filled him in on the behavior of the pups. Brian and the other rangers prepared to use more direct methods, such as firing cracker shells at the pups, to make them more wary of the road, people, and vehicles.

On Valentine's Day, 214 was back near the Agates with another of his brothers, 252. The Nez Perce wolves howled, and the Agates howled back. The Druids were to the west and they howled as well. I looked back at the Agate group and saw that one of the females, the recently collared 251, who like the other females in the pack had been born a Druid, was approaching 252. Within a few moments of meeting up, they got in a breeding tie. Then it got complicated. 21 led the Druids toward those wolves. It was not a good time for the Nez Perce male to be caught in a tie with 21's daughter. The pair broke apart and 252 raced off toward his brother. The female, now that she was free of the tie, rejoined the Agate group and they ran up Dead Puppy Hill. 21 and twelve other Druids were soon at the base of the hill. The six Agates looked down at them. Their alpha male, 113, did the wise thing and led his pack to the south at a run. I lost the group as they ran over a ridge. That was the only time we saw 251 and 252 together so their affair was short-lived.

Meanwhile another Nez Perce male, 215, who had previously been with the Agates for a while, was now hanging out with 105's Buffalo Fork pack, where he bred a gray female that most likely was another former Druid. That meant that after the confrontation with the Nez Perce wolves in late October, three of the males from the rival pack had mated with 21's daughters. In later years, I saw that pattern repeated many times. Two packs would come into conflict and fight. Young males and females from those families would often get together later and form new packs. It was the wolf version of what used to be called college mixers, where students from different schools would meet and pair off.

I tried to monitor 253 every day to see how he was fitting into the Druid pack's hierarchy. One day when the alpha pair were away from Lamar Valley, I saw him leading a group of six young wolves. It looked like he was now the pack's beta male. He still limped on his injured hind leg, but it didn't seem to hinder him significantly.

The two gray Druid pups that we were concerned about were in the Hitching Post parking lot in early March and one of them approached a park visitor within fifteen feet, a sign the pup had no fear of people. I sounded my horn. The pup moved off a short distance but did not seem too concerned. The pair walked down the road, passing within a few feet of stopped cars. They came back to the lot and one pup sniffed at a trash can.

I called Brian. The pups were still in the lot when he arrived. I gave him a quick report and we both ran at the pups and yelled loudly. Then Brian fired several cracker shells in the air. Those loud explosions scared the pups, and both ran away. One ran up to the den and the other went south, in the opposite direction. Later that pup tried to approach the road to return to the den area but turned around and ran off when he saw people walking down the road toward him. That seemed to indicate that the cracker shells had taught him to be wary of humans.

After all the excitement of the mating season, I thought back to an incident in February involving 21 and his pups. 21 was trying to lead his family to the east in Lamar Valley. None of the wolves followed him. He went back to them and later tried to lead east once more, but the younger wolves played together and ignored their father. 21 turned around again and went back to them. A gray pup ran over, did play bows to him, then jumped up and down in front of 21. Now in a playful mood, the big male imitated the little pup and also bounced up and down.

Most of the other wolves were chasing each other around a big boulder. 21 ran over and pursued a pup around the rock.

42 was also playing with the pups by that time. Two pups ran to 21 and each one jumped up on his back. Then a group of pups chased him around the boulder. He stopped, turned to those pups, and ran back and forth in front of them. One pup jumped up on him and they wrestled. 21 let the pup think he was an even match for him. After that 21 ran by a group of playing wolves like he wanted them to chase him. 42 and three pups obliged and ran after him.

When things slowed down, 21 once again tried to lead the pack east and this time they followed. He took them toward a large elk herd. The herd bunched up, then ran north. The wolves charged after them with the younger wolves outrunning the alphas. A cow cut away from the herd and instantly a gray and a black targeted her. They easily caught up and one wolf bit into a hind leg. A third wolf ran in and grabbed the cow's throat. The three young Druids pulled her down and within a few more seconds she was dead. 42 joined those wolves at that point, but the work was all done.

21 had repeatedly tried to get the pack to follow him east, but they ignored him. He went back and, after playing with the younger wolves, finally got them to follow him to where the elk herd was. Had he spotted those elk earlier and wanted to take the other wolves to them? When the others finally saw the elk, 21 let the younger wolves lead the chase and mount the fatal attack. He did nothing other than lead the wolves to a point where they could see the elk and let them do the rest. It was a successful training day.

As I thought through all the mating season developments, I realized that all three middle-aged Druid females—103, 105, and 106—had now helped to form new packs: Agate, Buffalo

Fork, and Geode. Each of the three packs had Druids as alpha females and there were younger Druid females in some of those groups, as well. 21's extended family was spreading out and dominating a large portion of Yellowstone.

10

The Battle of Hellroaring Creek

ON THE EVENING of March 13, I spotted the Geode wolves coming into the Hellroaring area: 106, the alpha male, and four other adults. Three Druid wolves howled from lower down the slope: the alphas and a black pup. I looked at the Geodes and saw them run together for a rally and group howl in response to the Druids. The Druids looked in their direction. 21 frequently led his pack into this area and must have considered it part of the Druid territory, but the newly formed Geode pack was now claiming it as their land.

The Geodes ran downhill at the Druids. Despite being outnumbered six to three, 21 immediately charged uphill to meet them. On the Geode side, their alpha male was out in front. 21 sped up and charged into the middle of the six rival wolves, just as he had done with the Nez Perce wolves the previous year. Intimidated by his decisiveness, the Geodes scattered. 21 chased a black, then the big gray alpha male. By then 42 had joined him. A large black male ran in and went at 42. She nimbly dodged his attack, then chased the wolf off. 21 ran over to help her.

As that black wolf fled, 21 saw the Geode alpha male. He tackled him, then repeatedly bit him as he squirmed on the ground. Other wolves ran in and all of them fought. I saw that the big gray was now up and attacking 21. He pinned 21 and bit him. But 21 jumped up and counterattacked. At that moment, the black male ran over and he and the big gray teamed up against 21. They chased him and the gray bit him on the rear end. 21 stopped and turned to confront the two males. The Geode alpha male, now afraid, ran off. 21 went after him, caught up, and bit him on the back.

Once again the Geode alpha male ran away with 21 pursuing him. 21 let the big gray go and ran over to 42 to check on her. I saw that the Geode alpha was likewise with his alpha female. 21 was not done yet. He chased off the black male, then went back after the Geode alpha pair. They all ran from 21.

From a military standpoint, 21 was in total command of the battle. He chased one wolf, then another, and none of the rival wolves could stand up to him. 42 stayed with him, while the Geodes scattered. Then all the wolves went out of sight. When they reappeared, 21 and 42 were surrounded by all six Geode adults. With the Druids so outnumbered, I expected that the six would gang up and attack the pair. But before the Geodes could act, 21 took command once again and charged at a gray. It ran off and the other five Geodes hesitated. Soon 21, with 42 by his side, had all six on the run.

Exhausted, the six Geode wolves stopped. 21 and 42 caught up with them and there was a moment when 106 stood next to the Druid alphas. The three former packmates seemed to call a truce that spread to the other Geode wolves and soon all six of them walked off uphill. I assumed the battle was over.

At that point, I noticed a seventh wolf in the Geode group, a young black. It was the Druid pup. Not understanding the seriousness of the situation, it probably thought this was a play session. The pup went to 106, who it would have known well, and they had a friendly reunion. Four other Geodes then attacked the pup. It got away and ran downhill toward 21, who was now alone. The big gray alpha male led the pursuit. When he reached 21, he attacked, and the two fought. The other Geodes ran in, and to protect his pup, 21 ran downhill with it, toward a group of Druids that were in the area but had so far not been involved in the fight. The rival wolves raced after 21 and the pup.

21 ran into some trees and the Geodes caught up with him there. All six attacked him, but since they were biting 21, that allowed the pup to get away. The situation was looking fatal for 21. Then the Geodes suddenly turned around and ran away. I looked back and saw that 42, 253, and the other Druids had seen what was happening to 21 and were all running in to rescue him. They chased the Geodes uphill.

Their enemies vanquished, the Druids stopped and came together for a rally and howl. I counted thirteen in the group. All were holding their tails high. 106 turned and looked downhill at her former pack, then resumed running uphill after the other Geode wolves. The Druids chased them once more with a gray pup in the lead. They soon stopped and 106 continued to lead her group in a retreat to the west.

I took a closer look at the Druid group and saw that 21 was not with them. Had the Geodes killed him when the six ganged up on him? The Druids ran back downhill, toward where 21 was last seen. I lost them in some trees. I frantically scanned that area and eventually spotted 42 calmly bedded down once more, looking like nothing of significance had happened. Standing beside her was 21. His left hip was bloody, but other than that he seemed fine. 21 and 42 howled and the rest of their family howled back. 42 got up and she and 21 walked over to the others, and they all had a big greeting.

While I was writing the details of that incident, I thought a lot about how 42, 253, and the other Druids were willing to risk their lives to rescue 21. He had put himself in jeopardy many times in the past when he protected them from rival wolves, and in this battle they chose to do the same for him, regardless of the danger to themselves. It seemed to me that the primary directive in 21's code of behavior could be summed up with a simple statement: protect your family, regardless of the cost. That is what 253 and the other younger wolves had seen 21 do for them, and that was their understanding of what they needed to do if they saw a family member under attack. I was especially impressed with 253 that day. Having a disability, he would have been justified in staying safely on the sidelines, but despite his injured leg, he was right in the middle of the rescue operation.

As I thought more about the incident, I recalled a poem Rudyard Kipling wrote in *The Second Jungle Book*. Entitled

"The Law for the Wolves," the verses list the proper behavior for pack members. The most intriguing line states, "For the strength of the pack is the wolf, and the strength of the wolf is the pack."

That profound sentence perfectly described what had happened that day. The strongest wolf and best fighter in the Druid pack was 21. There were many times over the years when the Druids relied on his strength to protect the family from rival wolf packs. But in this moment, 21 needed the combined strength of the pack to save his life. Inspired by what the Druid wolves did that day, I thought of another way to state the law of the wolves: "The wolf takes care of the pack, and the pack takes care of the wolf."

The Druids stayed in the Hellroaring area the next day, but the following day the alphas headed back to Lamar Valley. The other Druids remained at Hellroaring. One day I watched 253 traveling with the group. He was putting weight on his injured hind leg, but when he ran, instead of alternating one hind leg with the other, the usual pattern for a running wolf, he put both hind legs down at the same time. I guessed that it was too painful for that bad leg to hit the ground without the support of the other hind leg, so he improvised a unique style of running and could keep up with the pack when they were chasing prey.

One morning 253 and ten other young Druids chased a cow elk. The three lead wolves caught up with her and pulled her down at 9:32 a.m. 253 and the rest of the group ran in to help finish her off, but a nearby bison herd charged at them and the Druids had to back off. The bison surrounded the cow, not to protect her, but to sniff at the scent of blood on her. Some of the wolves slipped in between the bison and attacked the cow, but the bison moved in closer and the wolves had to retreat. Several bison continued to sniff at the bleeding elk.

The cow stood up but collapsed right away. She got up a second time, tried to run off, but fell down again. I saw her get up on her hind legs, but she seemed to have no control over her front legs. A bison approached, butted her in the rear end, and knocked her down. Now so many bison were surrounding the cow that I couldn't see her. All eleven wolves—four yearlings and seven pups—bedded down, waiting for the bison to leave.

Dave Mech was with me and we watched the cow slowly deteriorate from blood loss. A magpie landed on her and she did not react. As I checked on the nearby wolves, Dave watched the cow. At 11:35 a.m. he said she was dead. Ravens landed and pecked at her. The bison chased off the wolves when they tried to get to the carcass and even drove off the ravens. They seemed to be acting as protectors of the dead elk, but really were driving the wolves away from their own herd. At 1:13 p.m., the wolves finally managed to stand their ground when several bison tried to chase them away from the carcass. Nearly four hours after they pulled the cow down, they finally got to her carcass and fed.

The following day, March 21, I saw 21 and 42 trotting toward the Druids' traditional den area. At one point, 21 was well ahead of 42, who was slowed down by her advancing pregnancy. He stopped, bedded down, and waited for her. When she passed him, 21 got up and followed her. I lost them going into the trees near the den site. I thought about all the other times I had seen him follow her lead. The Druids often ignored 21 when he wanted to go in a particular direction, but they would follow 42 when she led, and he would join the line behind her. That was one of the clear examples that 42 was the leader of the pack, not 21. I think he understood that and had no problem with her being in charge. In the language of *Game of Thrones*, 21 willingly "bent the knee" to 42.

253 and the ten other young wolves spent several more days in the Hellroaring area and had a lot of success hunting elk. Signals from the Geodes indicated they were close by, but they never bothered 253's group. After their defeat by the Druids, they apparently wanted no part of another fight with those wolves. From a distance, 253 looked a lot like 21 and was rapidly growing bigger and more impressive looking. Perhaps the Geodes mistook him for his father.

Later genetic studies suggested the Geode alpha male was New Gray, the young wolf 21 allowed to join the Druids in August 2000. New Gray bred 106 in early 2001 and she had pups that spring. 21 drove off New Gray at the end of that year. In early 2002, 106 formed the Geode pack with four other gray wolves. Two of those grays were her offspring and likely sired by New Gray when 106 was still in the Druid pack. It seems he had started the pack with 106 and was the alpha male 21 fought at Hellroaring. We suspected that 21 originally let New Gray into his pack because the two wolves were related, and the later genetic studies did indeed suggest that New Gray was related to Rose Creek female 18, 21's sister.

11

Dens and Pups

On April 1, I got signals from 42 toward the ridge east of the Yellowstone Institute. That is where she denned in 1998 and 1999 when 40 was the alpha female and occupying the pack's main den. The ridge is forested, and we couldn't see her. No other Druid signals were in the area. Her signal got weaker for a while, then stronger, so we assumed she had gone in and out of her old den. Later I got 21's signal from that site. By April 5, the continued presence of 42 in the area indicated that she had chosen to den there again. 21 was the only Druid who visited her there regularly, as 253 and the ten other pack members were still in the Hellroaring area.

I had seen 21 and 42 mate twice on the last day of January so her due date was around April 4. Since 42 was the only adult female left in the pack unrelated to 21, she was the only one he had bred. A single litter would make his life as an alpha male much easier than it had been the previous two years. I saw 21 bedded down on a cliff above her den four days after her due date, watching over the site. Later he went down into the forest where her den was located. I saw her come out of the trees in the evening and she looked thin. That meant she must have had her pups. 21 joined her and the two of them went out for a walk. Soon after that, the other Druid adults and the previous year's pups, now yearlings, came to the site to visit 42. In mid-April, I got signals from 42 and the other Druids from the forest north of the Footbridge parking lot, the traditional den site for the pack. I wondered if 42 was thinking of moving her pups there.

On the last day of the month, a coyote researcher saw an uncollared black on the south side of Lamar Valley and noticed she was nursing. He lost her going up into the Druid den forest. That set off a lot of speculation. All the young adult females in the Druid pack were daughters of 21 and he had shown no interest in breeding them, the norm for closely related wolves. We had seen uncollared black Druid females that were temporarily in the newly forming packs or with lone males, and some were spotted breeding. That female must have been one of them.

By May 1, it looked like 42 had moved her pups to the main den. Thirteen Druids left the forest that day and traveled west. One of the black adults in the group had missing fur under her belly, a sign she was nursing. She looked like a two-year-old and had a lot of gray on her muzzle. She must have been the black the coyote researcher told me about.

Wolf watchers Mark and Carol Rickman had the first sighting of pups at the main Druid den on May 28. Three days later, five pups were seen up there: three blacks and two grays. I was in the Footbridge lot at the time. I took people in the lot out to the south and set up my spotting scope. I located some of the pups, then let everyone have a turn at the scope. Each person was ecstatic when they saw the pups. The little wolves wrestled with each other, then went into the trees. We assumed that all five belonged to the alpha female 42, but there was the question of the nursing black female we had seen visiting the Druid den forest. Were some of the pups hers, or did she have a den elsewhere?

In my field notes, I called her the grayish-black because she had a lot of gray streaking on her coat. I also noticed that she had marks on her hips in the shape of a parenthesis. Then I remembered that I had seen a black with those marks mate with one of the roving Nez Perce males, 214. I recalled that she had also been with Agate alpha male 113 during the mating season. In dog litters, there can be more than one sire, and research in Yellowstone later proved that to be true for wolves. Some of the pups in her litter could have been sired by 214 and others by 113. But where were her pups?

Doug Smith, Dan Graf, my co-worker that summer, and I hiked up on the ridge east of the Yellowstone Institute to look for the site that 42 had used prior to moving her pups to the main den. The den on the ridge was marked by a dead standing tree so we had no difficulty finding it. There were two openings through the roots, and both were too narrow for a person or bear to squeeze through. Doug extended a measuring stick into the den and found that it went back sixteen feet. It probably originally had been a coyote den.

I frequently went west to check on the new Agate pack. We knew the alpha male, 113, had bred 103 and now we were getting signals from her southeast of Tower Junction, toward Agate and Quartz Creeks, two small streams that flow down from Specimen Ridge into the Yellowstone River. We thought her den must be in that area. This was confirmed when later she was spotted with four pups in that area. The Agate alpha female was with her that day. She appeared to have no pups of her own and was probably there to help 103. They were likely sisters.

We often got signals from the Agate alpha male to the north of Tower Junction, near Garnet Hill. Eventually four pups were seen there. They were with an uncollared black adult female with a gray muzzle who seemed to be their mother. That meant that the Agate adults would have to travel back and forth between two den sites that season. The dens were only about five miles apart, but the wolves would have to swim the Yellowstone River during the time of high spring levels to get from one to the other.

I finally figured out the mystery involving the nursing black Druid female with the parenthesis marking when I spotted her behind a low ridge called Middle Foothill on the south side of the Chalcedony Creek rendezvous site. Three pups were with her: one black and two grays. After mating with 214 and spending time with the Agate alpha male, she must have returned to her family, the Druids, and had her pups in a new den site in those trees. So 21, like the Agate alpha male, had two den sites to tend to this season.

I watched the three Druid pups play with their mother and another black adult. 251, who divided her time between the Agates and the Druids, was visiting that day. The gray pup went to her and they touched nose to nose. The other two pups ran over and all three climbed around on her. She was gone the next day, and Dan later saw her with the four Agate pups at the Garnet Hill site.

As I monitored the Chalcedony site, I saw other Druids go up into the trees behind the Middle Foothill to where the den with the three pups would be. They alternated between visiting those pups and 21 and 42's five pups at the main den. Soon the pups to the south moved out into meadow in front of the Middle Foothill and explored the area, just as previous generations of Druid pups had done.

The Druids now numbered twenty-three if you included 251, who visited regularly, and 217, who had returned to the group after spending time with both the Geodes and the Agates. Eight of the pack members were pups, with five at the main den and three at the Chalcedony rendezvous site. The two Druid dens were just a few miles apart, and when wolves howled from the main den, the others at the rendezvous site would howl back.

217 seemed especially taken with the three pups at the rendezvous site. One day she led them down through the trees. When they lagged behind her, she stepped behind a tree and playfully jumped out at them as they walked by. Soon after that, she spotted a grizzly not far away. After taking the pups back up to the den, she repeatedly charged at the bear and bit him on the rear end several times. 21 had started to go that way to help but ended up bedding down when he saw that 217 was handling the grizzly on her own. Later she chased off a red fox that was near the pups. A few days after that, the mother wolf—the black with the parenthesis markings—drove a black bear that was approaching her pups up a

tree. The grizzly, fox, and black bear all could have killed the pups if the little wolves had been left unattended.

Two mornings later, 21 visited the three pups and their mother at the Chalcedony rendezvous site. After he left, the Agate alpha pair and two other Agate wolves showed up. I saw them sniffing at a spot where we had seen 21 just an hour earlier. Both Agate alphas had their tails raised. They would have gotten other Druid scents, including the female with the parenthesis marking who had been a temporary member of their pack during the mating season. That made me wonder if 113, the big Agate male, was trying to find her and see if she had pups that he might have sired.

The Agates moved toward the Middle Foothill. The mother wolf and 217, who had also been with the big Agate male earlier in the year, were bedded down at the base of the slope. They got up when they saw the approaching Agates and greeted them like old friends coming for a visit. The mother wolf playfully bumped up against the male's chest, then jumped up on his rear end. 217 did play bows to him. Neither Druid female showed any concern about his presence, proof that they knew him well and trusted him. Then 251 came out of the trees wagging her tail. The Agates went toward her and the other two Druid females followed. All seven wolves had a big greeting ceremony.

We hadn't seen the three pups for a while and figured they were up in the trees behind the foothill. The uncollared Agate alpha female went into the trees and the other wolves followed. The Druids still did not seem concerned. I lost the seven adults in the forest. They were out of sight for the next thirty-six minutes, then six of them reappeared and went out into the meadow. All of them, including the mother wolf, seemed relaxed and calm, like they had met up with the pups and had a friendly interaction with them. The missing wolf was Druid female 251, who was likely still with the pups.

The Agate alphas did some scent marking, then went back up behind the foothill. The mother wolf followed and came back with two of the pups. After sniffing around where the Agates had been, mother and pups went back up toward the den. No wolves came back into view during the next hour, but I continued to get signals from 113 and 251 from the trees. Don Robertson, a good wolf spotter, volunteered to watch the area while I did other things. I got back four hours later and got a signal from 251, but nothing from the Agate male. I joined Don and he showed me 251 bedded down on the slope at Middle Foothill. Earlier he had seen the mother wolf and two pups behind that hill. The Agates, it seemed, had left. They had likely taken the trail up Chalcedony Creek. That would get them to the top of Specimen Ridge, and from there they could go west to their den site.

The peaceful visit of the Agate alpha pair to that Druid den site was a fascinating event to witness. We thought the Agate alpha female was a former Druid, so the three pups would have been related to her. Since the mother wolf had been with the Agates in the mating season, 113 may have thought those pups were his. I recalled that when he was a yearling, he liked to play with the pups in his family, so the big male likely did that with these pups.

In late evening, 21, 42, and three other Druids traveled from the main den to the Chalcedony rendezvous site and met up with 251. In the area where we had seen the Agate male do raised-leg urinations, 21 did scent marking. He seemed unconcerned about getting the scent of the other alpha male. 21 returned to 42 and they bedded down, as did the other Druids. Don and I stayed till almost 10:00 p.m. and saw the adults going up into the trees to where the three pups should be.

I got back to that area at 5:00 a.m. the next morning. The signals from 42 and 253 were coming from the main den, and 21 was to the south. I went up on a hill to look at the Chalcedony rendezvous site and saw 113 and the other three Agates arriving there from the west. A black Druid female and two of the pups were on the foothill. The pups headed into the trees, acting like everything was normal. The Agates ran to the Druid female and greeted her, then went up into the forest. The mother wolf came into sight and followed their route. Soon the four Agates reappeared from the trees and the mother wolf and another Druid female were with them. They all socialized. I saw the mother flirt with 113. Two more Druid females ran in and joined the group. All eight—four Agates and four Druid females—interacted in a friendly manner.

At 6:00 a.m., I spotted 21 and three male yearlings moving in from the east. Two of the Druid females noticed 21 and went to him. With the approach of the Druid males, the dynamics of the group at the rendezvous site suddenly changed. I looked west and saw the dark gray, the second-ranking Agate male, running west. When 113 spotted 21, he ran northeast. 21's group chased him for a while, then turned and went after the other male. More Druids joined 21. The Agate alpha changed direction and raced toward his packmate. 21 was now targeting 113 and closing in on him, but then 21 stopped and let him go, seemingly satisfied that the other male was leaving. The big male joined up with two Agate females and continued west, but frequently looked back toward 21.

I lost the Agates to the west just after 7:00 a.m. By then the Druids were calmly bedded down at the rendezvous site. Dan Graf drove west and later radioed me that the Agates went up and over Specimen Ridge at 8:00 a.m. That would take them back to their main den. I concentrated on watching 21 and saw him playing with a male yearling. He jumped on the young male and they sparred with their jaws. Later 21 and the other wolves ran up into the trees to the south to visit the pups.

The following day, 21 was back at the Chalcedony rendezvous site. We once again saw only two pups: a black and a gray. There had been a second gray, but we had not seen it in a while. There were grizzlies, black bears, cougars, and coyotes in the area, and any one of them could have grabbed the missing pup. I didn't think any of the Agates had harmed it. The number of pups in the main den was also down, from five to four.

A few days later, I saw two gray pups with 42 and other adults south of the main den. Both pups had black streaking along their backs, a marking that comes in gradually on gray pups as they age. One of the pups ran in circles as it tried to catch its tail. I lost those wolves going west toward the river. Other observers told me they saw 42 pick up a stick as she approached the water. I soon spotted 42 on the other side of the river and noticed that she was still carrying the stick.

One of the gray pups and a black pup were following her. She must have used the stick to entice them to swim across the river with her. The other gray pup was still east of the river.

In the evening, we saw 42's two pups at the rendezvous site. A short distance away were two smaller pups: a black and a gray. The pups' size and the absence of streaking on the gray's back confirmed these pups were younger than 42's. We heard that 42's third pup had crossed the road and gone back to the main den. Doug flew over and confirmed there were now two pups at the main den and four pups at the rendezvous site. On his next tracking flight, Doug looked down at the rendezvous site and saw pups going in and out of the burrows north of the Middle Foothill, tunnels that must have originally been dug by coyotes.

On July 4, all six pups were at the rendezvous site, and that evening I had my first sighting of the combined litters. The four older pups were much larger than the two younger ones. A big gray pup chased a small gray, which tucked its tail between its legs. It ran to a nearby yearling and stood under its chest. The yearling put a paw over the pup's shoulder, looking like he was comforting or protecting it. Later two of the big pups stood over the little pup and it snapped at them defensively. When that pup ran off, a big pup chased it. But the young mother came over and got between her pup and the bigger one.

Both smaller pups turned out to be females. The black was more aggressive than her gray sister, and I saw her chasing one of the big pups. The small gray pup gradually learned to fit in, and I soon saw her chasing big pups around the rendezvous site. She once wrestled two of the larger pups at the same time. They worked together to pull the smaller one down, but she jumped right up and continued the match. Later she chased one of them, grabbed a hind leg, and wrestled it to the ground. After that the two small pups teamed up and harassed the bigger pups. Once when a big and a small pup were sparring face to face, the other small pup sneaked up and bit the big pup on the rear end.

A black wolf arrived at the rendezvous site and regurgitated six times to the pups. That was a record number of regurgitations. Then 42 and 217 arrived and both regurgitated three times. Later 21 came in, but the pups were too stuffed to bother him for food. A young black female who had been watching over the pups was nearby, so he regurgitated to her, a reward for her taking on the babysitting duties.

253 shared his father's enjoyment of playing with pups. One evening the pups ran to their bedded older brother. He chased them, sparred with his jaws, and later rolled on his back and wiggled his paws in the air as the pups stood over him.

On July 24, I drove to the Antelope Creek area. From there I had a good view of the south side of Specimen Ridge. I spotted six Druids near the Agate den, including 21 and 42. The Agate alpha male, 113, was watching them intently from a distance. He and other Agates had visited the Druid den site earlier in the summer, and now 21 and the Druids were returning the favor, checking out the den where former Druid 103 had had her pups.

The next morning, I found 21 and the Druids back in Lamar Valley. They went to the rendezvous site, then walked up behind the Middle Foothill looking for the pups. They must have been moved by their mother for the adults soon reappeared, then marched up into the trees at Chalcedony Creek, and I lost them. The pack had a route in the drainage that would take them up toward the top of Specimen Ridge, where their pups were probably stashed.

For the next month, the pack alternated between their two rendezvous sites, the upper one at Opal Creek and the lower one at Chalcedony Creek, as they searched for elk to hunt and patrolled their territory.

Deb Guernsey from the Wolf Project office flew on August 6 and got signals from 21 and 42 at the Opal Creek rendezvous site. The alphas and twelve other adults were back at Chalcedony Creek two days later. Three pups were with them, all of them females: two grays and one black. That turned out to be the count of surviving pups.

12

Wolves and Ravens

On August 23, a Wolf Project flight found sixteen Druids back up at the Opal Creek rendezvous site: thirteen adults and the three pups. Since I had once hiked up toward that area and watched the Druids from a distance, I imagined 21 and 42 bedded down on the hill in the meadow. From that rise, they could watch over the pups as they played nearby. On the flight, Doug Smith got a mortality signal from Druid male 254 up the Lamar River, but thick trees prevented him from seeing anything on the ground. Doug and other staff hiked in and found the wolf's remains among boulders at the base of a cliff. They concluded that he had fallen to his death. I later saw wolves confronting elk on cliff tops. 254 had probably done the same thing and either got too close to the edge and slipped or was kicked by an elk and fell.

The three Druid pups were back at the Chalcedony Creek rendezvous site the following day. 42 wrestled with one and played tug-of-war with another. She was well over seven years old now, about sixty in human years, but she still had a playful spirit. Later she ran to 21 and affectionately licked his face. The two had been together now for nearly five years, about the same length of time as the entire life-span of an average Yellowstone wolf.

21, like 42, was getting old. Despite his advancing age, he often acted like a much younger wolf. One day he caught a vole and tossed it in the air several times. After the last throw, he touched the tiny rodent with his nose, then immediately jerked his head away. The vole must have bitten him. After that he picked it up, carried it around, then dropped the vole and playfully pawed at it. At that point, the black pup came over, hoping to get the toy from his father. Wanting to prolong the game, 21 grabbed the vole, carried it off, bedded down, and resumed playing with it. The pup joined him, then lay down a few inches from the vole and 21's face. Several other wolves approached and 21 made the mistake of turning his head to look at them. The pup, waiting for just such a moment, darted in, grabbed the vole, trotted off, and tossed it into the air. Then she ate it. After that 21 played with the big gray female pup. They wrestled and he made it seem like they were evenly matched. She snapped at him and bit at the fur on his face. When she ran off, the big alpha male romped after her. They continued to wrestle and spar, and later 21 played the ambush game with her.

By late August, it was getting harder to find the Druids. Although they occasionally showed up at the Chalcedony

Creek rendezvous site, they spent most of their time out of sight up on Specimen Ridge.

Elena West, a Wolf Project staff member, and I hiked up the Lamar River in early September. We had seen the Druids investigate an active beaver lodge there the previous June. The unusually high water in the spring had destroyed most of the structure. The family had made a bad choice of location for the lodge and abandoned the site. There was no sign that any beavers were still in the area.

On the way back, I noticed how well the willows were doing along the river and at nearby Soda Butte Creek, where there was a new beaver colony, perhaps started by the family that had lived upriver. Beavers rely on willows close to their lodges as their primary food supply, and the willows were rebounding after years of heavy browsing by elk. In a meadow near Pebble Creek Campground, the willows were growing especially wide as well as tall, up to fifteen feet.

I regularly hiked up Crystal Creek to where one of the wolf acclimation pens had been built. I led nature hikes up there for several summers after the 1995 wolf reintroduction and showed people how elk had killed thousands of aspen shoots by overbrowsing them. Now the young aspens in that area were twenty feet tall and as thick as a bamboo forest. That month I talked with a vegetation researcher who had 113 study plots in the local area. He found that aspen shoots were an average of twelve centimeters higher in areas where wolves were common compared with plots without wolves. He interpreted that to mean that elk were spending less time browsing on aspen shoots in areas where wolves were frequently encountered.

More importantly there were fewer elk in the park. Wolves, along with mountain lions, were the main predators of elk when Yellowstone was set aside in 1872. The early rangers killed off the last of the original wolves in 1926 and shot the last lions in the 1930s. In the absence of those predators, elk multiplied to an overpopulation level that did great damage to the vegetation they fed on, especially willows, aspens, and cottonwoods. Now, with the reintroduction of wolves, predation by grizzlies and cougars, and increased human hunting of elk adjacent to the park, elk numbers were in better balance with food supplies. With fewer elk in the park, willows, aspens, and cottonwoods were recovering, and that benefited other plant-eating animals such as bison and moose, as well as beavers. The restored vegetation also provided additional habitat for songbirds.

Throughout the summer, I continued to monitor the Geode pack. Now the pups had the stamina to travel full-time with the adults. I saw the family at Tower Junction, in Little America, and on Specimen Ridge. Those areas were also used by the Agate wolves, but we didn't know of any encounters between the two related groups. By the end of September, the Geodes traveled as far east as the Amethyst Creek drainage in Lamar Valley, but went no farther, probably because they wanted to avoid encountering the Druids. That month I also had a few sightings of 105's small group, the Buffalo Fork pack.

We took some Native American Elders from the Kiowa Tribe out to Lamar, and they got to see the Druids at the Chalcedony rendezvous site. One of the Elders told us a story about a Kiowa man who had traveled through Yellowstone in the old days before it was a park. He had heard wolves howling and later turned his experience into a song and interpretive dance for his people. The Elder then sang the song for us. Native peoples such as the Kiowas had an especially close relationship with wolves in this region for thousands of years. They studied how wolves hunted and organized their packs, and applied what they learned to their own lives and tribes. This was the second time I had shown wolves to a group of Native people, most of whom had never seen a wild wolf, and I felt that it was a special privilege to do so.

Later in the fall, I tallied up how the packs were doing with their pups. The Buffalo Fork pack did not seem to have any. The pup count in the Geode pack was down from ten to three. There had been eight pups seen at the two Agate den sites and now there were only four. The three surviving Druid pups were doing well, but the count of adults had dropped to nine from a high of sixteen earlier in the year. One had died in the fall from the cliff, and the other missing yearlings had likely dispersed to seek out mates.

253, the Druid with the injured leg, would occasionally disappear for a few days on a walkabout, probably to find a female to pair off with, then come back. He was two and a half years old, the same age as 21 when he left the Rose Creek pack and joined the Druids in November 1997.

In October I was asked by the Make-A-Wish Foundation to take a teenage girl and her family out to look for wolves. Amanda had watched Bob Landis's recent television documentaries on the Druids and her wish was to see 21 and his family. We found the Druids at the Chalcedony rendezvous site and watched them for three hours. It was a humbling feeling to help make her wish come true. Later Make-A-Wish contacted me again and we helped another young girl see wolves. Her trip was particularly special because two Native American friends of mine, John Potter and Scott Frazier, were in the park to conduct a wolf blessing ceremony and they invited the girl and her family to join them.

Previously I had taken out a retired Japanese wildlife professor who was dying of cancer. Wolves had been native to his country, but they had all been killed off before he was born. His last wish was to see wolves in the wild and we helped him see the Druids. Another man was in a wheelchair and we pushed him high enough up a hill so he could see wolves.

Anyone in my position would have taken these people out to see wolves, but I was especially motivated to help because of a moving episode I had witnessed when 21 was still in the Rose Creek pack and helping to care for the young pups. On that occasion, after bringing food to the main group of pups, he noticed one that was apart from the others and seemed to have health problems. 21 went to that pup and spent some time with it, an act of empathy that must have cheered it up.

I always agreed to help any sick or disabled people who wanted to see wolves, but I also helped everyone else who came to me. Some were Oscar-winning movie stars, television personalities, models, singers, and US senators. One was a former British prime minister, and another was a billionaire, but 99.99 percent were regular people. Once a married couple came up to me and said that years earlier I had shown their kids wolves. They added that those kids were now grown up with children of their own. When they got together on holidays, the family would always talk about that day when they saw wolves in Yellowstone. I once tried to estimate how many people I had helped see wolves in the park and figured it might be around 100,000.

I was often asked how close I had ever been to a wolf pack. I replied by telling people that when I watched wolves, they were an average of a mile away. If I accidently came across wolves while I was hiking and they were nearby, I would back away. Wolves normally avoid people and we wanted to maintain that behavior. Otherwise, when they left the park, they might think it was safe to be near humans and get shot, as several Yellowstone wolves have been in recent years. A mile was also about the average distance between the wolves and the groups of visitors I helped.

Early one morning, I saw two black Druids at the Chalcedony rendezvous site. One was very dark and the other was grayish black with a light U-shaped mark on her chest. A herd of five elk with a calf came into the area. The nearest black ran at them and singled out the calf. There was only one cow in the group, and she veered away from the calf, indicating it was not hers. The calf was soon separated from the adults by the wolf but was easily outrunning it. The black showed no sign of giving up.

At the one-minute mark of the chase, a raven flew in from the west, reached the wolf, and kept pace with it. It must have spotted the pursuit from a distance and come over to see if it might result in some food. The bird called out several times. The wolf was now gaining on the calf. Speeding up, it grabbed a hind leg. The calf bucked up and down, trying to shake off the wolf. Then it kicked back with its other hind leg. The raven circled overhead, watching the interaction, and continued to squawk. Three other ravens flew in. The calf was still struggling to break free, but the wolf held on. The calf ran in tight circles as it bucked and kicked. The other wolf ran in and bit the calf's throat. It soon was dead.

The grayish-black wolf who had initiated the pursuit was now feeding on the calf. It looked like the mother of the pups born at the rendezvous site, the one with the parenthesis marking on her hips. From that time on, due to her prominent chest markings, we called her U Black. Another black Druid female had even more gray on her coat than U Black and we referred to her as Half Black. The Wolf Project tries to refer to uncollared wolves by a designation other than a human name, so we look for some distinctive physical trait.

That first raven had spotted the chase right after it started, flown over the wolf and calf, then called out. The three other birds heard those calls and understood their meaning. Perhaps they were the original bird's mate and offspring. The four ravens were now on the ground a few feet from the wolves, looking for a chance to steal bits from the carcass.

Ravens nearly always find an animal that has died of natural causes before wolves, and a wolf can learn how to profit from that. Years earlier I had spotted a Druid yearling bedded down north of the road in Lamar Valley. He was staring off to the south, where there was a lot of raven activity. The wolf got up, crossed the road and the Lamar River, and went straight to where he saw the ravens. I put my scope on the site and saw a cow elk carcass, untouched by wolves or bears. The ravens had spotted it from the air and their presence tipped off the wolf. Since the carcass was intact, the ravens couldn't peck through the cow's thick hide. Once the yearling tore into it, the meat would be accessible to the birds. They needed a wolf to open up the carcass for them.

Another time I saw a flock of fifty ravens circling the Druid wolves as they traveled, looking like they were escorting the pack. Then the birds glided off and landed four hundred yards away from the wolves. I saw something sticking out of the snow there but could not tell what it was. The lead wolves paused to stare at the ravens, then ran in their direction. The flock flew up when the pack arrived and circled over them. Two wolves went to where the ravens had been, sniffed around, then jointly tugged on the object protruding from the deep snow. A few moments later, they pulled out an elk carcass that had been completely covered by recent snowstorms. The ravens knew about that site but couldn't yank the carcass out by themselves. Did the ravens deliberately circle the pack, then fly to that site, thinking the wolves would be curious, come over, and pull the carcass out so that the flock could scavenge on it? I think they did exactly that.

In those two cases, wolves found a new carcass thanks to ravens tipping them off. But there were many times when ravens found a fresh kill when they saw wolves at the site. They would then steal meat from the pack. Ravens were chased off when they got too close, but the wolves would generally tolerate their presence. That may be partly because pups get used to the frequent presence of ravens looking for a free meal around den sites. For however long there have been ravens and wolves, they have had a mutually beneficial relationship.

In the early years of the wolf reintroduction in Yellowstone, I recalled reading about a captive raven that lived to be sixty-nine years old. That suggested a wild bird could also live that long. Perhaps an elderly raven might have been around when the wolves were brought back to the park in 1995, sixty-nine years after the last of the original wolves had been killed. On seeing the new wolves, the old bird might have remembered its early years and how it fed alongside wolf packs when they made kills. The first time it saw one of the new packs on a fresh carcass, the raven would land and steal meat from the site. Other ravens, too young to have known about wolves, would catch on and swarm to the carcass. Chris Wilmers, who studied raven activity at wolf kills for the Wolf Project, once counted 163 ravens at a carcass.

Another Wolf Project researcher, Dan Stahler, did an innovative raven–wolf study in Yellowstone as part of his master's degree program at the University of Vermont under the supervision of famed raven expert Dr. Bernd Heinrich, author of the book *Ravens in Winter*.

As an experiment, Dan put out road-killed animals in places where no wolves were present. He recorded how often ravens found one of the sites during the first hour and how they reacted to the carcass when they found it. The birds discovered the carcasses within an hour only 36 percent of the time compared with 100 percent of the time when wolves made kills. In an even more important finding, he discovered that when ravens found a carcass that had no wolves on it, they circled a few times and then flew off. None came back to feed during that first hour. In contrast, if there were wolves at a new carcass, ravens landed right away and tried to feed.

In the research paper Dan and Bernd published on that experiment, they suggested that ravens have a fear response to new carcasses when wolves are not on the scene. They called that neophobia, fear of new things that might be dangerous. If ravens see wolves at a new carcass, that apparently reassures them the situation is safe, so they land and feed.

How would that behavior pattern develop? In past times, people poisoned carcasses to kill off wolves and coyotes that fed on them. The poison killed birds as well, especially ravens. Perhaps earlier generations of ravens saw fellow birds dying after eating at those sites and learned to avoid them. But they also learned to watch for wolves at carcasses and if ravens saw them feeding without any ill effects, that indicated they could safely eat there as well. The wolves were like the food tasters monarchs used to employ to sample all their food before it was passed on to them to eat.

Years ago when I was researching a talk on ravens, I found an account that put an intriguing spin on the subject of poisoned carcasses. A man in Canada looked out of his cabin window and spotted a raven feeding on a new carcass. Later, when he took a second look, he saw a large flock of ravens circling the area. He shifted his view to the carcass and saw the original raven stretched out on its back, motionless. The entire flock flew off without ever landing, apparently concluding that the bird at the carcass had died from eating poisoned meat. Impressed at the intelligence and ability of the ravens to analyze the situation, the man took another look at the site. Now that the big flock was out of sight, the dead raven got up and resumed feeding. It was just a trick to avoid sharing the meat. When other ravens flew by, it played dead again.

While I was working in Alaska at Denali National Park, I read a lot of Native stories from that region about a character named Raven, who was always trying to trick people and other animals, usually out of food. Because of that behavior, he was known as Trickster. Experts in that field of study estimate that the stories originated in Siberia twenty thousand years ago and were brought to Alaska over the Bering Land Bridge. They may be the oldest known stories still being told today. When I read the account of the raven that fooled the other birds, I could see that ravens' behavior in the wild paralleled the behavior of Raven Trickster in those stories.

13

Invasion and a Separate Peace

IN LATE NOVEMBER of 2002, the Mollie's wolves came up to Lamar Valley from their territory to the south. Their ancestors, the Crystal Creek pack, had been driven out of Lamar by the Druid wolves in the spring of 1996 prior to 21 joining the pack.

That morning the valley was fogged in. We heard howling from what seemed to be three separate groups of wolves. Two of them were south of the road, and the third was north of the road. We could see the group north of the road: two of the three surviving Druid pups with a black adult.

It began to clear to the south and we spotted twelve Mollie's wolves: ten adults and two pups. I then saw the second Druid group with seven wolves, including the alphas and the pack's third surviving pup. Despite the presence of rival wolves in their territory, the pack didn't seem concerned.

21's group howled and the Mollie's howled back. The Mollie's wolves alternated between looking at the main group of seven Druids when they howled and the other three Druids to the north when they howled back. The larger Druid group turned toward the Mollie's when they were howling. I had not seen the Mollie's wolves for years and did not know the pack members. A big gray wolf with a raised tail seemed to be the alpha male. They trotted off and I lost them as they traveled west. I swung my scope to check on the main Druid group and saw that the wolves were sleeping.

Two hours after the Mollie's left, the Druids got up, found the scent trail of the Mollie's, and excitedly sniffed around. I could hear the Mollie's howling in the distance. The Druids headed toward the sound and howled back. I soon spotted the twelve Mollie's running toward the seven Druids. 42 had turned around and was now leading her family north at a run, away from the Mollie's. 21 was right behind her, in a position to protect 42 if the Mollie's caught up with them. 42 was likely planning to join the adult and two pups on the other side of the road. If the Druids reunited, they would add up to ten, about the same number as the other pack.

At that point, 21 stopped, looked back, and howled. 42 followed his example. The Mollie's briefly ran toward the Druids, then veered away from them. 21 stood in plain sight staring at the other wolves. He was standing directly between the twelve Mollie's and 42. They would have to go through him to get to her. That made me think of the battle the Druids had with the Nez Perce wolves when 42 had been separated from 21 and she got injured. I think 21 had learned from that and was now staying right with 42, ready to protect her.

The twelve Mollie's wolves were now running south, away from 21. I looked back at him. He was standing all alone, calmly watching the other wolves retreating. I heard a howl to the north and saw 42 on the far side of the road. She was looking straight at 21. He turned and headed her way. When he joined the other Druids, they all howled, laying claim to their territory. Fourteen minutes later, I checked for signals from the Mollie's and got none. I studied 21 and was amazed to see that he was calm and unstressed. It looked like his quiet confidence and determination to protect his mate and family by standing between them and the rival wolves had intimidated the larger Mollie's pack into leaving the area.

There were times when I used what I had learned from 21 in personal situations. One day when I was in one of the parking lots in Lamar Valley, a car pulled in at high speed. The driver got out and she seemed frightened. She spotted me in my ranger uniform and ran over. The woman told me that she and her young daughter had just been involved in a road-rage incident. They were driving along when a car pulled up behind them. The male driver repeatedly honked his horn and yelled at them. She was worried he was going to come for them in this lot.

I called a law enforcement ranger on my Park Service radio, briefed him on the incident, and gave him our location. Right after that, the man swerved into the lot and jumped out of his car. I told the mother and daughter to get behind me. The man rushed over. Remembering how 21 had stood between 42 and the Mollie's wolves, I stayed put and calmly looked at the man.

As he came at us, he glared at the mother and girl, then noticed me. A moment later, he slowed down. Then he looked around and backed off. The patrol ranger came along at that time. I pointed out the man and the ranger dealt with him. Just as 21 had watched his adopted father, wolf 8, and applied what he learned from him to his own life, I did the same thing from watching 21. To me, he was the gold standard of dependability and courage.

Two days after the incident between the two packs, the Mollie's were north of the road. This time there were thirteen of them. The Druids were on the south side of the valley. They heard the Mollie's howling and looked toward the sound.

217 was in the Druid group. She had been born to 21 and 42 in 2000, and was currently a part-time member of the family, spending some of her time with the Agates. Now two and a half years old, she needed to find an unrelated male so she could start her own pack, as so many of her sisters had done. We stopped seeing 217 with the Druids soon after that. I saw her traveling alone on December 7 at the west end of Lamar Valley. There was howling to the north of her, across the road. 217 went that way and soon we saw her with two wolves we didn't recognize, a black and a gray. All three were excitedly greeting each other. The black did a raised-leg urination and 217 marked his site right away. The gray male also did a scent mark there. 217 acted in a playful manner with the new wolves.

On the day before Christmas, I found 261, a gray male from the Mollie's, with 217 and her two companions near a carcass at Slough Creek. The three males were friendly to each other, and that indicated they likely were brothers. 217 flirted with all three and they responded in kind. Those four wolves would soon be designated the Slough Creek pack. 217 and 261 were the alpha pair. It was an alliance of a Druid female with several Mollie's males, two packs that had been enemies since the spring of 1996, almost seven years before. Two days later, a black female joined them. She and 217 got along well together, so the new arrival was probably a Druid.

This was the fourth new pack that dispersing Druid females helped to start in 2002: Agate, Geode, Buffalo Fork, and now Slough Creek. All four groups laid claim to sections of the superterritory that 21 had taken over in the latter part of 2000. Instead of one superpack, there were now five normal-sized packs that split that vast territory into subsections.

At the end of 2002, the Agate pack had six adults, including alpha male 113, the alpha female, who was likely a Druid, former Druid females 103 and 251, and the black who had the litter of pups at Garnet Hill. Four of the original eight Agate pups survived. The Geode wolves, with Druid 106 as the alpha female, had six adults and three surviving pups. The pack started by Druid 105, the Buffalo Fork wolves, had four adults and no pups.

Throughout the year, I had regularly hiked up South Butte to watch the Leopold wolves, a pack I had been studying since it was started in early 1996 by yearlings 2 and 7. I had seen the pair raise many litters of pups. I had been especially impressed by how playful and affectionate they had been throughout their long association with each other.

Alpha female 7 had died during the spring denning season. When a Wolf Project crew examined her, they determined that she had fought with other wolves and succumbed to her injuries. She had been seen with eight pups and was still nursing at the time of her death. Alpha male 2 and the twelve other Leopold adults worked together to keep 7's pups alive and well in the months after her death. An uncollared black, likely a daughter of 2 and 7, became the new alpha female. By the end of the year, some adults had dispersed, but all the pups had survived.

Matt Metz, who was on the Leopold winter study crew, told me that 2 had left the pack in late November and was now traveling with five blacks and two grays. All the other females in the Leopold pack would be daughters of 2, meaning he would not breed with them. A black female and a gray female in the group he was traveling with were possible mates for him. I saw 2 put a front paw on the gray, then lay his head over her back, a sign of his interest in her.

He never did get to start a new pack. On the last day of the year, we got a mortality signal from 2 in the Hellroaring area. 106's signal was in the area. A Wolf Project crew found 2's remains and concluded that he had been attacked by rival wolves, most likely the Geode pack, and died of his wounds. As there were no wolves unrelated to 2 and 7 in the Leopold pack, they mated only with each other during their long life together and had at least thirty-nine pups. Twenty-nine of them survived to be yearlings, and two of their daughters founded packs of their own. The Leopold pack lived on after the pair's death, and two of their sons would be destined to play major roles in the Yellowstone wolf story.

I was especially saddened to hear that both 2 and 7 had died fighting rival packs, but that is the reality for wolves in the park. It is the most common cause of death for adults and I accept that fact. When you study wolves intensely for years, you get to know charismatic individuals you greatly admire. Eventually you will witness or hear about their deaths. I think wolves love the lives they have been given and accept whatever difficulties come with that life without feeling sorry for themselves. I try to do the same.

14

253's Incredible Journey

So much was happening with all those packs during the last few months of the year that I have held off telling a story that overshadowed everything else that was happening in the Yellowstone wolf population.

253 was last seen with the Druids on October 17. That evening I had watched him lead his family to the Chalcedony Creek rendezvous site. I got his signal and the other Druids' signals there early the next morning. I counted ten wolves but did not see 253. His signal faded and soon was gone. He must have taken off by himself and gone up and over Specimen Ridge.

Then he disappeared. We couldn't get his signal anywhere in the park, even during tracking flights. A lone dispersing wolf with a permanently injured hind leg did not have a good chance of surviving on his own. If he had stayed home, he would have been the successor to 21 when he passed away, meaning 253 could have been the next Druid alpha male. But the urge to find an unrelated female to pair off and have pups with must have outweighed that option.

On the last day of November, a coyote trapper in the Wasatch Mountains, twenty-five miles northeast of Salt Lake City, went out to check his traps and found a large male black wolf caught in one by the front right paw. The trapper saw a second set of wolf tracks in the area, meaning the male had been traveling with a companion, probably a female. The man tied the wolf's legs together, took him out of the trap, and drove him to the local Utah Department of Natural Resources warden, who put the wolf in a kennel, then fed and watered him. His paw was treated by a veterinarian.

Mike Jimenez, a biologist with the U.S. Fish and Wildlife Service, took possession of the animal. The wolf had a radio collar and the frequency matched the one assigned to 253. He was the first confirmed wolf in Utah in over seventy years. The place where he had been found was about two hundred miles from Lamar Valley. Mike drove 253 northeast and turned him loose in Grand Teton National Park, near the South Entrance of Yellowstone, on December 2. I later talked to Mike about 253 and he told me that the wolf had been very docile to handle.

The straight-line distance from that release site to Lamar Valley was sixty miles. Mike monitored 253's signal and located him east of Yellowstone Lake on December 10, about forty-five miles south of Lamar. I got his signal in Lamar Valley early on December 20, sixty-five days after I had last seen him with the Druids. I spotted the pack near Amethyst

Creek and counted nine wolves. One was limping. It was 253. He must have joined his family during the night. It had taken him eighteen days to get back home from Grand Teton.

When the pack started to travel, I saw that 253 was last in line, slowed down by the old injury on his left hind leg and by the newly hurt right front paw. He kept that paw off the ground at times, but when he held his left hind leg up, he had to use that front paw, regardless of the pain. As the pack continued on, 253 had a hard time keeping up. At one point, he lay down to take weight off those legs, but got up after a few minutes and resumed following the rest of the wolves. The Druids traveled for hours that day and 253 managed to stay with them.

A few days later, when the Druids bedded down after a traveling session, I saw 253 licking that front paw. I noticed that 21 often chose to bed down next to 253 when the pack rested. On the day before Christmas, 253 was second in line as 21 led the pack at a fast pace. Three days later, 253 didn't limp on that front paw when he ran to greet 21 and 42. After that he only occasionally kept it off the ground when traveling.

A week after 253 was caught, another wolf stepped into a coyote trap in Utah, but pulled out of it and ran off as a game warden approached. Ed Bangs of the U.S. Fish and Wildlife Service and the leader of Northern Rocky Mountain Wolf Recovery told a reporter that he thought it could have been the wolf who had been with 253 when he was trapped. Ed felt 253 had been traveling with a female and the pair had been looking for a territory to settle down in.

Brent Israelsen from the *Salt Lake Tribune* interviewed me about 253's return to Lamar Valley. In his published story, he quoted me as saying, "He was well known before his trip to Utah. There has been a lot of sympathy generated by his story and condition. All that multiplied 253 to a higher level of notoriety. He's a real celebrity now." The headline for the article read "Beloved Wolf, 253, Running with Original Pack."

In another story, the same reporter interviewed Doug Smith. Doug spoke about how, after 253 injured his hind leg and returned to his family, he captured the public's imagination for his diligence in tending pups, hunting elk, and defending the pack's den from bears: "He did more than some of the more healthy wolves." Doug told the reporter how impressed he was by 253's ability to travel such a long distance to Utah with one bad leg, then return to the Druids with two injured ones: "They don't have a concept of difficulty or pain. They don't say to themselves, 'Gee, I got a limp so I can't go very far today.' They just do it." Those lines truly summed up 253's character.

I thought a lot about that second wolf in Utah. The two wolves had found each other and probably were getting ready to have a family, then 253 stepped in that trap. When I worked in Denali National Park, I was friends with wolf biologist Gordon Haber. During his research in that area, he documented several cases where a wolf would get caught in a trap and its mate would stay with it, only running off when the trapper came on the scene.

That sense of loyalty sometimes extended beyond death. In his book, *Among Wolves*, Gordon wrote of a male wolf who stayed with his female for two weeks as she struggled to get out of a trap. The trapper then arrived, killed the helpless wolf, and hauled her away on a snowmobile. The following day, Gordon saw the male mournfully howling for his mate at her capture site. The male returned to the site repeatedly for weeks, vainly trying to find his female. His trips outside the protection of Denali had a tragic ending when a hunter shot and killed him.

The presence of the second set of tracks near 253 when he was in the trap suggested that he had found a companion with a similar sense of loyalty. I would like to think that she later found a mate as good as 253.

PART IV

2003

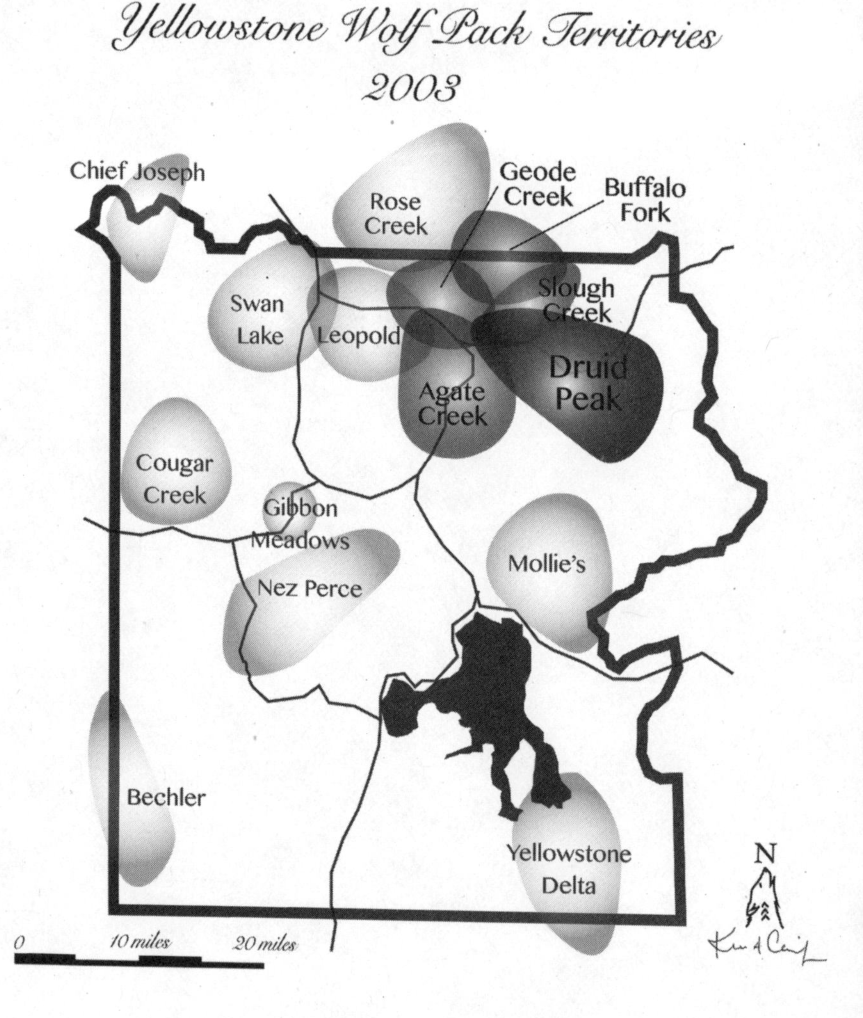
Yellowstone Wolf Pack Territories
2003
Chief Joseph
Rose Creek
Geode Creek
Buffalo Fork
Swan Lake
Leopold
Slough Creek
Druid Peak
Agate Creek
Cougar Creek
Gibbon Meadows
Nez Perce
Mollie's
Bechler
Yellowstone Delta
N
0
10 miles
20 miles

PACKS INCREASE AND decrease in size over the course of a calendar year. These charts show the main pack members in any given year. M=male and F=female. An asterisk (*) indicates a female thought to have denned. The pack of origin for wolves joining from other packs is indicated in parentheses the first time a wolf is introduced. Squares indicate adults and yearlings. Circles indicate pups.

Druid Peak Pack

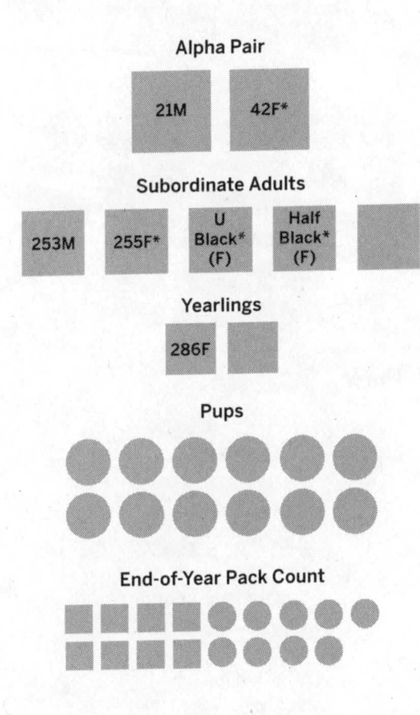

Agate Creek Pack

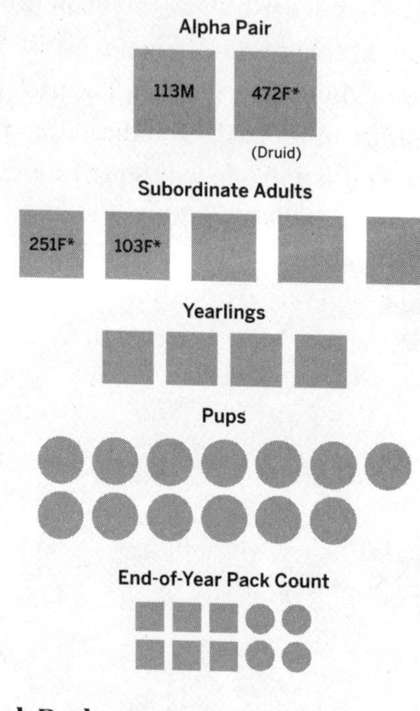

Slough Creek Pack

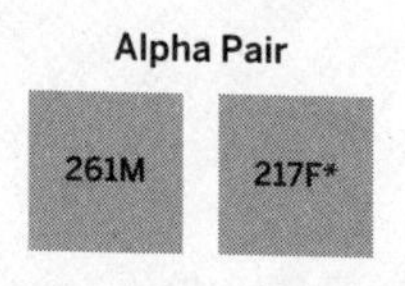

Subordinate Adults

Count varied throughout the year

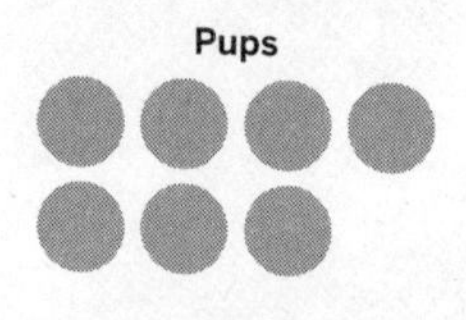

End-of-Year Pack Count

Geode Creek Pack

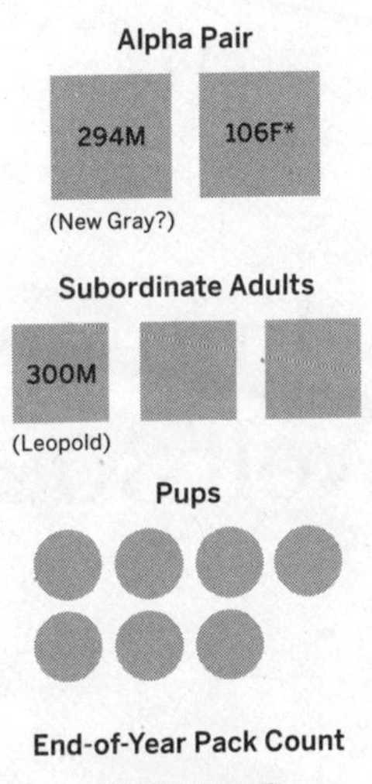

Unattached Males

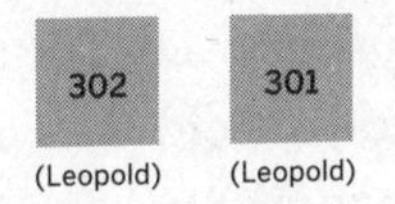

Druid Females Seen With 302

15

Enter Wolf 302

IN EARLY JANUARY, the Druids usually numbered nine and were down to two pups from a high count of eight. The two remaining pups were female, one black and one gray. This was a stark contrast to 2000 when twenty of the twenty-one Druid pups survived. The seven adults were the alphas 21 and 42, two-year-old 253, a black male yearling, and female yearlings 255, U Black, and Half Black. The yearlings were almost two years old now. That was the age when some of them might leave the pack to seek out mates and start their own families.

Doug Smith started radio collaring for the season. He darted 217, now the alpha female of the Slough Creek pack, and replaced her nonfunctioning collar with a new one. Then he darted the Druid black pup. She was designated wolf 286. The crew took a blood sample from the pup and it showed that Nez Perce male 214 was her father. He was the wolf I had seen mate with U Black before she returned to the Druids to have her pups.

Later in the day, Doug darted three wolves in the Swan Lake pack, a group founded by a daughter of Leopold wolves 2 and 7 that had staked out territory near Mammoth Hot Springs. My friend Annie Graham was in the park at that time. I had stayed with Annie and her husband, Bob, when I was invited to give a grizzly bear talk at the Houston Zoo two years earlier. I called Doug on the radio and he said we could hike out and help with processing the wolves. We joined the crew and watched as they examined the animals for health issues, took blood samples, then fitted each one with a radio collar. The tranquilizer drug used on wolves works quickly to immobilize them. They are semiconscious while being processed and keep their eyes open.

The other pack members were nearby, and they howled, trying to contact their three missing members. As soon as everything was done, we all hiked out. Over the years, I have been involved with dozens of wolf-capture operations and the newly collared wolves are nearly always back with their packs by the next morning. They find them by hearing the others howling.

On January 17, it was 1 Fahrenheit (–17 Celsius) when I left my cabin at 6:39 a.m. I spotted all nine Druids east of their den. They howled and looked southwest. I turned my scope that way and saw a lone wolf. It was a husky young male with a sleek black coat. The animal was strikingly good looking. He could have won the Best in Show award at the Westminster Dog Show. We were about to enter the wolf breeding season and it was easy to figure out what this guy was up to. He must have gotten the scent of 21's three adult daughters. All he had to do was to lure the sisters away from their father, run off with them, and start his own pack.

He looked toward the Druids, then trotted in their direction. The Druids disappeared into a forest. The male sniffed around where I had lost the pack. Soon Half Black came out of the trees. She went to the newcomer wagging her tail. Then she sniffed him and did play bows. This looked exactly like the footage Bob Landis took in late 1997 when 42 first met 21. The black male looked past Half Black and ran off with his tail tucked between his legs. 21 and 42 charged past the female at the intruder, but the new male was far to the south by then. 21 found his scent trail and ran along it. Half Black and the other seven Druids followed.

I found the black wolf across the road. He was bedded down and casually looked back at the other wolves. 21 was glaring at him. U Black was next to her father. As she watched the intruder, she howled. The alphas joined in. Then 21 bedded down, seemingly satisfied that this unknown wolf had fled when challenged.

But the black wolf stayed where he was. U Black slipped away and trotted toward him. When she got close, she wagged her tail. They touched noses, then she did a play bow. He sniffed the female and I heard her make a whining sound. Then he ran off. She followed immediately and they ran west, side by side, away from her parents. I looked east and saw 21 and 253 charging after them. 253 ran without a limp on either injured leg. 21 was in the lead and he sped past U Black. The intruder, desperate to escape from 21, ran right through the middle of a big bison herd and kept on going. 21 stopped, probably figuring that this was the end of the incident.

When 21 turned around and trotted back toward his family, his daughter's suitor made what seemed like a fatal mistake: he ran at 21, looking like he was going to challenge him to a fight. At that moment, 21 looked back, saw the other male charging at him, and instantly ran toward him with his hackles up. I had never seen 21 look so intimidating.

The intruder lost his nerve and ran off, but was too slow. 21 caught up and instantly pinned him. The other male did not struggle or fight back; he just lay there submissively. 21 had him at his mercy and could easily have torn out his throat, but 21 did not attack; he just held the black down. After that demonstration of power, 21 let his opponent go, like a medieval knight showing mercy to a defeated opponent who had yielded to him. The new male ran off and 21 went back to U Black and 253.

I began to relax a bit after witnessing the intense incident, but then saw it wasn't yet over. The newcomer, despite his defeat, was walking back, straight toward the three Druids. He stopped, then watched them. 21, probably thinking this wolf would not be stupid enough to cause any more problems, walked off and 253 followed. U Black, however, went the other way, right toward the black male, once again wagging her tail. They got together and flirted. The new male went west and she followed. I no longer had 21 or the other Druids in sight.

I lost the pair to the west. Park volunteer Ray Rathmell was at a different angle and he called on the radio to say 21 had just run back and chased off the male. I saw the outsider moving off to the south. 21 was north of him and watching. U Black was back with 21. Later she slipped away from her father and once again met up with the black male. She tried to flirt with him, but he looked past her, then ran off at top speed. I turned and saw 21 charging at him. He soon lost sight of the intruder and went back toward 42. U Black and the new male made one more attempt to meet up before 21 drove him off yet again.

That male was later darted and collared and given the number 302. Genetic testing determined that he was the son of Leopold alphas 2 and 7. He had been born in 2000, so was two years and eight months old. Since his mother was 21's half sister, 302 would be 21's nephew. Even though the two wolves were closely related, 21 and 302 were opposites in personality, as I would soon learn.

Later in the day, Bob Landis saw U Black join up with 302 once more. 21 chased and caught him, then held him down. 302 went totally submissive. 21 let the male go, as he had done earlier, and 302 walked off. U Black followed her new suitor. Bob's report, added to what I had seen that morning, indicated that 302 had figured out how to take advantage of 21. If he went limp and didn't fight back when pinned, that appeared to end 21's aggression. 21, when he was a yearling, had witnessed how the male who had adopted and raised him, wolf 8, had defeated the original Druid alpha male in combat, then let him go. 21, apparently following the pattern he learned from 8, had never killed a defeated male, and 302 was now using that to his advantage.

I later saw 302 near the pack. The Druid male yearling chased him, and 302 ran off from the younger and smaller wolf. They ended up in a confrontation and 302 went into a crouch, tucked his tail between his legs, and walked off without putting up a fight. Satisfied that he had dominated the intruder, the yearling went back to his family.

The next day, we saw 302 with Half Black and another black Druid. The day after that, all three of 21's daughters and the male yearling were with 302. The yearling appeared to be dominant to 302. That evening the five wolves got up on a rock outcrop with the setting sun behind them. One howled and the others joined in.

The following morning, the two pups joined them and the group numbered seven. I went back there later and heard that the Druid alphas and 253 had arrived, greeted the six Druids already there, then attacked 302. 42 grabbed 302 and shook her head as she held him in her jaws. If it had been 21 who caught 302, his trick of going submissive would probably have worked again, but it wasn't going to fool 42. The male jumped up and ran off. He stayed away from the Druids for the next few days.

I next saw 302 on January 24. He was with U Black and two males we didn't recognize. Since the three wolves were getting along well, the two new ones were probably 302's brothers from the Leopold pack. All three were trying to court U Black. She seemed to favor 302.

21 was concentrating on something else during that time. On the evening of the twenty-sixth, 42 stood in front of him, averting her tail. He sniffed her and soon the pair was in a breeding tie. It lasted twenty minutes.

The next day, all nine Druids were at the west end of Lamar Valley, and we wondered if that meant 21's three daughters were over their infatuation with 302. 302 was a few miles away hanging out with another black male near Slough Creek. Both wolves ended up being collared on the same day, and the black male was designated as wolf 301. His DNA showed that he was a younger brother to 302.

Leopold yearling 301 was dominant to 302 just as the Druid male yearling was. 302 seemed willing to accept a subordinate position to other males, even younger ones, without standing up for himself. He behaved like a man with a premonition that he was going to die in a fight and therefore never wanted to get into one.

21 and 42 mated again on the twenty-ninth at 7:30 a.m. They got in another tie at 12:33 p.m. That was the third time I had seen them mate this season. While 21 was otherwise occupied, U Black was with 302 and his brother.

In early February, we often saw the two Leopold brothers with U Black and Half Black. 255 sometimes joined them. 302 tied with Half Black on the fourth and with 255 on the tenth and again the next day. 255 went back to the Druids soon after her second mating with 302.

The Wolf Project did more radio collaring on February 12 and I got to see Doug dart two Geode wolves in the Hellroaring area. He first tried to get the Geode alpha female, 106, but she ran into some trees, so he had to give up on her. Spotting a black male in the group, Doug hit him with a dart and the wolf bedded down within a minute. 106 had come back into sight by then and Doug got her as well. If you come back as a wolf in your next life and want to avoid being collared by a researcher, hide in a forest, but stay there until the helicopter leaves.

One afternoon I saw an incident that aptly demonstrated how different 302 and 21 were. 302 and his brother were on a fresh elk kill with U Black. 302 fed on a piece of meat away from the carcass. He looked over at U Black and saw her pull off a strip of meat. He ran over, pounced on her, and stole the meat. She calmly went back to the carcass and resumed feeding. I had never seen 21 steal food from a female. Despite that rude behavior, 21's three daughters still flocked to 302. One woman told me that 302 was like the bad boy who rode a motorcycle to high school, the one many of the girls had a crush on.

U Black was still with 302 and his brother the next day, February 15. Half Black was there as well. I had been keeping track of the female hierarchy. U Black was the dominant sister. She would pin 255, and later 255 would go to Half Black and pin her.

255 and Half Black were back with the Druids on the nineteenth. Soon after they returned, 253 initiated a play session with the black pup. That set off Half Black and she romped after the gray pup. 21 saw that, ran ahead, dropped down in the sage, waited for the two to catch up, then jumped up from his hiding spot and ran ahead of them again. When no one chased him, 21 ran back to them, then rushed off once more. Another day 253 played the chasing game with the black pup. During the chase, he put weight on his injured hind leg and ignored the pain. His right front paw, the one caught in the trap, seemed to be on the mend.

The snow was deep in Lamar Valley by late February. One day I saw 21 leading his family through a particularly difficult snowfield. At first, he stayed on top of the thin crust, but it didn't hold his weight and he started breaking through into the soft powder below. Then he had to struggle to make any forward progress. The black pup was right behind him. When 21 paused to rest, she moved around him and took over the lead. Now the big father wolf followed in the tracks of his daughter. When she got tired, the gray pup went out in front.

It was minus 39 Fahrenheit (–39 Celsius) on the morning of February 24 and my car wouldn't start. I had a portable battery and cables that I kept inside my cabin and they got the car going. The low temperature for the park that morning was minus 51 (–46). It was warmer in the valley, just minus 40 (–40). All three Druid females were back with the Leopold brothers on the twenty-fifth.

In 217's Slough Creek pack, alpha male 261, from the Mollie's pack, seemed to have a personality like 21's. I saw him play with two other males who were lower ranking to him. He wrestled with one and twice allowed the other wolf to pin him. Then 261 played with a black male. The other wolf chased 261 and pulled his tail when he caught up with him. I checked 261's genetic report and found that he was a grandson of wolf 8, the wolf that raised 21.

On March 10, Half Black and U Black were with 302 and his brother west of Slough Creek. The four wolves looked north, then ran off in the opposite direction. I then saw the Geode wolves charging at them with the alpha male in the lead and 106 right behind him. I soon lost sight of all the wolves. Ten minutes later, I picked up U Black and Half Black on top of a cliff. 302 must have run right past them to save himself, for he was at the base of the cliff, below his two females.

The next morning, I found 302 back with the two females. At one point, 302 looked north, then sped off to the south. I scanned to the north and saw the Geode wolves charging in. Half Black and U Black ran off and eventually caught up with 302. The trio managed to escape from the other pack, but once again 302's response when his females were in danger was to run off and save himself.

I had seen 302 mate with both 255 and Half Black. 255 had apparently figured out that 302 didn't have the makings of a proper alpha male and had gone home to the Druids. A few days later, U Black also rejoined her family. Dan Stahler did a tracking flight on March 20 and saw that 302 was back with the Leopold pack. That suggested that Half Black, the last of the three Druid females in his group, had also dumped him.

I spoke to Doug that day and he told me about a fight he had seen between the Mollie's wolves and a big bison bull in Pelican Valley. During the battle, one wolf was kicked and thrown fifteen feet through the air. Another was hooked by a horn and tossed a similar distance. A female wolf was already injured by the time Doug arrived on the scene, and she went into the brush to recuperate. Later she was found there dead with one of her legs broken. Remembering my conversation with Dave Mech in 2001, I wondered if she might also have suffered internal injuries. The two blacks that were tossed in the air recovered. The uninjured wolves fought on and eventually killed the bull.

Hunting bison was a dangerous endeavor, but when elk left due to deep winter snow in Pelican Valley, bison were usually the only available prey for the Mollie's. Big bison bulls weigh as much as 2,000 pounds and usually stand their ground when wolves approach. Therefore, being large is an advantage for wolves hunting bison, and the pack needed as many big males as possible. The Mollie's males tended to be the largest wolves in the park. In contrast, elk, the preferred prey for most packs, run from wolves, which means there is an advantage to having a lot of fast wolves in a pack. It is easier to be fast if you are not too heavy.

I once saw a wolf chase a bison at Slough Creek. As the wolf came in from behind, the bison kicked back and struck its pursuer full on. The wolf flew through the air, did a complete 360, then crashed to the ground. The wolf got right up and continued chasing the bison. Another time I saw five wolves confront a cow bison and her calf. The cow charged at a male wolf and he must have gotten a horn under his belly for he was thrown up into the air and twisted end over end. He had the misfortune to land on the cow's head. She flung him off and he crashed to the ground. Amazingly, the wolf got right up and seemed unhurt, like a superhero in a Marvel movie. Proof of that took place a few minutes later when he walked over to a nearby female and tried to flirt with her.

March 21 was the eighth anniversary of the release of the original three Alberta packs from their acclimation pens. Three days later, on March 24, we saw 302 walking east down the road in Lamar Valley. He veered off to the north and I lost him going into the Druid den forest. I had seen the Druid alphas and other pack members west of there. Half Black was not with them and might have been at the den. If so, 302 would meet up with her there. Matt Metz, who was on the Leopold winter study crew, later told me that 302 had been with the Leopold wolves the previous day. On March 26, Matt spotted 302 back with the Leopold pack. That meant that 302 took a three-day, fifty-mile round trip to visit the Druid den site. Did he do that to check on the Druid females he had mated with?

Near the end of March, I spotted the nine Druids. 42 looked pregnant, as did the three younger females. On the first of April, I saw 302 bedded down south of the road near the Footbridge parking lot. I heard he had come down from the Druid den forest earlier. That suggested he had tried to visit the three females he had been with during the breeding season. I then spotted those females and 253 at a nearby bull elk carcass. 302 did not approach them. Instead, he sniffed around two old Druid kill sites, found an elk leg, and buried it in a snowdrift. I had thought he would try to reunite with the females, but perhaps he was afraid of 253.

The three sisters were based at the main Druid den, along with 42, an indication that all four Druid females were planning to den there. Around that time, I saw the Agate alpha female and she looked pregnant, as did 251, who was now back with the Agates. Toward the end of February, I had seen 103 mate with 113. The black alpha female ran over and pounced on 103 during the tie, but she was too late to prevent the act. 113 likely mated with the alpha female at some other time. All this meant that 113 could have three litters to care for that spring. 217, alpha female of the new Slough Creek pack, and 106 in the Geode group were also pregnant.

302 was back in Lamar Valley on the eleventh. He sniffed around the Chalcedony Creek rendezvous site. Later I got his signal on the north side of the road near the Druid den. He found an old kill site and scavenged there. Four days later, I saw him going up to the Druid den forest. I got signals from 42 and 255 there, but not 21 or 253. Since the males were gone, 302 likely met up with some of the Druid females. He

was back with the Leopold wolves the next day. That pattern continued for some time. 302 would travel from the Leopold territory to the Druid den area, then return home again.

The battery on 21's radio collar failed on April 16. Most wolf-collar batteries last four years, but this collar had been on him for only fourteen months. The working collars in the Druids were down to four wolves: 42, 253, 255, and the black pup, 286.

I once saw a radio collar save a Druid female wolf's life. She was on her back being attacked by a big male from a rival pack. He reached down and bit into her throat with all of the 1,500-pound pressure in his jaws. But nothing happened to her, for he was biting into the battery pack attached to her collar. Her little brother ran in at that moment and attacked her assailant, and that allowed her to escape.

On April 23, I saw missing fur under the belly of Half Black, a sign she was nursing pups. Later that day, I saw the same thing on the underside of 42 and noticed her nipples were distended. In early May, 255 and U Black had the same signs. All three sisters had been with 302 during the mating season and I had seen him in ties with 255 and Half Black. He likely bred U Black as well. That meant 21 would be raising his own pups that spring and pups sired by 302.

16

Raising Pups

DOUG SMITH DID a flight in late April and saw the Slough Creek alphas at a den site east of Slough Creek Campground. He got a mortality signal from the Geode alpha male and saw his body at the base of a cliff. Two days later, Doug Smith and Dan Stahler hiked out to the site and examined his remains. Doug told me it appeared he had been killed by a kick from an elk or bison. There was blood in his mouth and blunt trauma on his left side. Apparently, he had been kicked so hard that he had been thrown through the air and probably died on landing or soon after. After the death of the Geodes' founding male, wolf 300 assumed the alpha position. He was a Leopold wolf and one of 302's brothers.

An elk died of natural causes south of the Druid den forest and a grizzly found the site before the wolves. In the evening, the Druid alphas, 253, and U Black must have gotten the scent of the carcass for they went directly there. The four wolves ran in and surrounded the bear. 21 bit the bear on the rear end twice. As it chased 21, the others fed on the elk. When the grizzly stopped chasing him, 21 went back to it with his hackles up and snarled. That angered the bear and it went after him again. While 21 harassed the bear, the two nursing females and his son fed.

The bear repeatedly charged at 21 and tried to swat him with a front paw, a blow that could have killed him, but the wolf was too agile and always dodged the strike. In frustration, the grizzly went back to the carcass and sat on top of it, looking like he was daring the Druids to try something. The four wolves surrounded it. The bear swatted at 21 once more. Then, as 21 grabbed a piece of meat from one end, 253 did the same from the opposite end. The bear swatted at one wolf, then the other, but both males dodged the blows. 21 teased the bear into chasing him once more and that gave the females more time to get in and feed.

The next morning, seven of the adult Druids went on a hunt. They spotted an elk herd and 253 chased a cow downhill. He ran all out without favoring either of his injured legs. 21 was also running at his top speed but could not keep up with his son. 253 caught up with the cow, ran alongside her, then lunged for her neck. She leaped up to avoid the attack and jumped over his back. 253 caught up with her again and bit into the side of her throat. 21 grabbed the right side of her belly. Then 42 ran in and grabbed the elk's rear end. The cow collapsed but managed to get up. 21 let go of her stomach and bit her upper throat. The two males then worked together to pull her down. 255 and Half Black ran in and soon all the Druids were feeding on the kill. I later saw 253 limping on his left hind leg. I guessed that on hunts he ignored the pain in that leg so that he could run at full speed. When he was not chasing prey animals, he reverted to limping because it was less painful.

The following day, I saw the Slough Creek alpha female, 217, show the same fierce determination 21, her father, had shown in standing up to the grizzly bear. I had spotted a female mountain lion on a fresh elk kill at Slough Creek. The Slough Creek alphas and a gray male came into the area from the direction of their den site. They must have gotten the scent of the carcass, for they ran directly toward it. The alpha male bypassed the lion, who had stepped away from the site when it saw the wolves, and went to the carcass to feed. But 217 had a different plan. When the lion ran off, the alpha female chased her at top speed. The wolf and lion seemed to be about the same size and weight. During the first part of the chase, the lion outran the wolf, but cougars don't have the long-distance endurance that wolves possess and soon 217 was gaining on her.

I looked ahead and saw a lone tree four hundred yards away from the carcass site. The cougar was aiming directly for it, but 217 was now right behind her and still gaining. When she was ten feet from that tree, the lion leaped at a forty-five-degree angle, grabbed the trunk with her claws, and speedily climbed up. 217 arrived a few moments later. She jumped onto a big boulder near the tree and looked up at the lion, who was now casually watching her, confident the wolf could not reach her. Knowing there was no chance of getting her opponent, 217 went back to the carcass and fed with the two other wolves. Thirty minutes later, the cat jumped down and disappeared into a thick forest. 217's den was a short distance away, and that was probably the reason for her aggression against the lion.

On May 9, I met up with Matt Metz at Hellroaring Overlook and he showed me the Geodes' new den site to the west, in a crevice at the base of a cliff. We saw five dark pups there that morning. A gray yearling came in and gave some meat to alpha female 106. The next morning, I saw her nurse her pups. Soon after that, I saw seven pups at that site. Doug did a flight on May 15 and found that 106 had moved her pups from the cliff base to a new den that was out of sight to us.

On that flight, Doug got signals from 251 in the Elk Creek area. She seemed to be denning there. She had finally decided to be an Agate after spending time the previous year with her birth family, the Druids. The Agate alpha female was collared and assigned the number 472. Genetic testing confirmed that she was indeed a Druid, born to 21 and 40 in 2000.

In mid-May, I spotted the Slough wolves at a coyote den located just west of Slough Creek. The coyotes had been harassing 217's family in recent weeks and it looked like she was retaliating. 217 dug at the den entrance as the alpha male and three other wolves kept the coyotes from attacking her. The site was on a slope and the den tunnel went horizontally into the hillside. 217 repeatedly went all the way into the den, then backed out. The alpha male came over and looked into the opening. It was now big enough that he could join 217. She came out with a limp coyote pup in her mouth, dropped it, and went back into the den. Soon she came out with a second pup. After that the Sloughs left the area.

A coyote researcher monitored the den for some time after that incident and told me that none of the coyote adults went back to the site. The abandoned den would be used by the Slough wolves in 2005 as a site to have their own pups. It is not unusual for wolves to repurpose a coyote den, and it wasn't the first time I had seen a female wolf killing the pups of a coyote family that was harassing her. The Sloughs used the site for three denning seasons and later three other packs denned there.

As in past denning seasons, 21 worked the hardest to bring food to the new pups at the Druid den. On May 18, I saw him coming in from the south. His belly was full to the bursting point and he carried an elk leg in his mouth. He crossed the road and I lost him going into the den forest.

Wayne Kendall found Agate female 251's den in the Elk Creek area and counted three pups. He showed me the site and we monitored it from across the road. We had a high count of five black pups. Genetic analysis showed that 251 was another of 21 and 42's daughters. On May 30, wolf watchers Mark and Carol Rickman, who spent a lot of time on Dead Puppy Hill, had the first sighting of a pup at the Druid den. Dan Stahler flew over the site and called down to say he saw seven black pups. He counted eight black pups at the Agate pack's main den where their alpha female, 472, was based. There were seven pups at the Leopold den. Another season of pup rearing was getting underway for the wolves of Yellowstone.

In late May, I bought a small cabin in Silver Gate and moved out of my rental cabin. It was the first home I had ever owned. I still live there.

At that time, 302 was mostly at Blacktail Plateau with his family, the Leopolds. In early June, he traveled the twenty-five miles to Lamar Valley and tried to cross the road to the den, but drivers stopped their cars and he turned back. He did make it across eight hours later and was seen entering the den forest. 302 had mated with at least two Druid females so he probably met up with pups he had sired. Two days later, he was back with his own pack.

That spring there was a research project on elk calf predation. It was conducted by University of Minnesota graduate student Shannon Barber, who had Dave Mech for an adviser. Her crew radio-collared thirty-one calves. As of early June, eight had been killed by bears, two by wolves, and one by a coyote. Another calf died from drowning. That data showed wolf predation was accounting for only a small fraction of calf deaths, just under 17 percent. Bears killed four times more elk calves than wolves did.

I met a game warden from the Utah Division of Wildlife Resources in Lamar Valley, and he told me that he had cared for 253 after the wolf was caught in the coyote trap. 253 was in sight at the time and as we watched him, the man told me his agency had placed armed guards around the wolf to protect him from people who might have tried to kill him. Then they handed him over to U.S. Fish and Wildlife biologist Mike Jimenez, who released him near the South Entrance of Yellowstone. The Utah warden didn't say anything more as we watched 253, but I sensed he was pleased to see how well the wolf was doing.

On June 6, Dan did a tracking flight and got a mortality signal from 105 in the upper part of Hellroaring Creek.

On the previous flight, she had been in that area with the Rose Creek pack nearby. Her group, the Buffalo Fork pack, had taken over part of the original Rose Creek territory, and there had likely been a confrontation between the two packs. Dan and Matt hiked out there and found 105. Her injuries indicated she had been killed by other wolves. She was lactating, but her pups would be eating meat by that stage in their development so there was hope they could survive with the help of the other pack members.

On June 12, I completed three years of going out early each morning to look for wolves and study them. That was 1,095 days in a row. A few days after that, I saw 21 and 253 supervising pups at the Druid den. The two big males had taken on babysitting duty while the mother wolves took a break.

During Dan's June 26 flight, he got a mortality signal from Agate female 251 west of Dunraven Pass, southeast of her den site. We hoped that her collar had dropped off and that she was all right. Matt and another Wolf Project staff person hiked out there and found her remains. A grizzly or lion had likely killed her, fed on her, then buried her remains. No meat was left on her body and the hide was torn apart. A puncture wound from a large tooth was found on her head. This was the second mother wolf killed this summer, and once again we wondered if the pups would survive.

On Dan's next fight, he saw twelve pups at the Druid den. The following day, I went up on Dead Puppy Hill and watched the area. Yearling 286 came out of the trees with six black pups and three gray pups. One of the grays and two of the blacks were substantially larger than the other pups. We were still not exactly sure which of the four breeding females had produced the pups and where the pups had been born, there at the Druid den or at the other den site south of the road.

One day I was talking to a man and his wife as we watched the Druid wolves. They had just driven through a nearby western state and met a local man who had spoken at length about his dislike of wolves. He said that each wolf kills and eats twenty-seven elk per week.

Doug Smith told me that our research in Yellowstone estimates a wolf needs fourteen to twenty-two elk per year, a mixture of adults and calves. Those elk could be ones the wolves killed, ones that other predators killed, or elk that died of natural causes. Yellowstone has a large bison population and adults die of natural causes on a regular basis. A bison bull or cow would be equivalent to three or more adult elk. That story of a wolf killing twenty-seven elk every week was an example of the misinformation being spread around about wolves. The best way I could counteract that was to show people wolves in the park and tell their stories.

On July 1, I saw six of the Druid adults, including 21 and 42, leaving the den area. After they crossed the road to the south, eight pups appeared on the slope above the road. They were wary of following the route of the older wolves. A few cars came by and the pups ran back into the den forest. The adults went up the Chalcedony Creek drainage an hour later. Two hours after that, Doug flew over the area and got the Druid signals at the Opal Creek rendezvous site, high up on Specimen Ridge. Early the next morning, I got all those signals back at the den. The adults must have gone out on a hunt while leaving the pups at the den.

I spotted Druid female 255 and seven pups just south of the road at 5:22 a.m. on July 3. They were on the north side of Soda Butte Creek. The pups were making little squealing noises, an indication they were stressed. 255 crossed the creek to the south. The pups refused to follow, so she came back. One of the gray pups slipped into the water and the fast current swept it away. 255 plunged into the creek and swam beside the pup. That cut the force of the current on the pup. Both wolves swam to a gravel bar in the middle of the stream and climbed out, then easily crossed through shallow water to the south bank. They traveled west as the other six pups watched them, still afraid to get in the creek.

255 turned around and went back to the other pups. The first gray pup continued west on its own. After greeting the pups, 255 waded back into the water, but none of them followed. She returned and tried to pick up one of the pups, intending to swim across with it in her mouth, but it seemed too heavy for her.

She went to the far bank and looked back at the pups. One black stepped into the water, then stopped. Another moved toward the creek, but a sibling playfully grabbed its tail and held it back. The six pups called out in distress. Several swam and waded across to 255, but one got swept away by the current. 255 ran after it on the south bank, reached down, grabbed the pup, and dragged it out of the water. The rest of the pups got safely across right after that. 255 then led south as she followed the scent trail of the first gray pup that crossed the creek. The other six pups followed her in single file.

Soon 255 reached the east bank of the Lamar River, a much wider and deeper body of water than the creek. She waded in and swam toward the west bank. None of the pups followed. 255 returned to them and led them farther south to a shallow ford. She waded and swam that section, but the pups still wouldn't go into the water.

I noticed that among the seven pups there were three different sizes: small, medium, and large. That suggested there were at least three different litters in this subgroup. Five other pups had been seen up at the main den and another litter might have been born at the rendezvous site den.

255 tried many more times to get the pups across the river and repeatedly failed. The pups ran back to the north and reached Soda Butte Creek. A small black pup started to swim north but struggled in the swift-flowing water. 255 jumped in and waded alongside the pup, blocking the strong current from sweeping it away. She tried to grab it in her mouth, but the determined pup ignored her and swam the rest of the way to the opposite bank. Seeing that pup succeed, the other pups followed its route across. All of them crossed the road to the north and returned to the den forest. But the first gray pup was still out there, somewhere south of the creek, by itself. The next morning, I saw that missing gray pup at the Chalcedony Creek rendezvous site with two black adults. It had to have swum the Lamar River to get there. Five other adults, including the alphas and 255, joined them. 42 and other adults played with the pup.

I drove west and saw 302 walking down the road. This had become a habit for him. It was easier for him to use the road as he traveled the twenty-five miles from his family at Blacktail Plateau to Lamar Valley. He soon veered off to the north and headed toward the Druid den forest. I heard the pups howling from there. I got a report that Half Black and five pups were visible at the den. I saw 302 going into the den forest, close to where Half Black had been seen.

Twenty-two minutes later, I spotted 21 going up to the den forest from the south, directly toward 302's likely location. I checked and got a strong signal from 302 that way. I visualized 21 spotting 302 and charging at him. I did another check seventeen minutes later and no longer got 302. Later I did get him to the east. It seemed that he had once again run away from the Druid alpha.

That evening I found 302 at the Chalcedony rendezvous site. The lone gray pup was following him around, suggesting he knew and trusted 302. I saw the black female yearling 286 bedded down nearby. 302 went to her wagging his tail. He sniffed her and she snapped at him. After that he followed the pup around. 302 smelled it, then they bedded down together.

I then heard that 21, 42, Half Black, and the gray yearling had just come down from the den and were heading toward the rendezvous site. Half Black and the gray yearling got there first and had a friendly reunion with 302. I scanned to the east and spotted 21 charging toward 302. He saw 21 and took off. 21 and the two females ran after 302. He got away and went back to the Leopold territory. I had questioned 302's sense of duty and responsibility in the past, but now saw that he was making an effort to visit the pups he had likely sired in the Druid pack.

The next morning, July 5, there were six pups (two blacks and four grays) at the rendezvous site. The adults must have brought the new ones there during the night. Six adults were with them. 42, 286, and Half Black soon went back across the road to the main den and I saw them there with six other pups (four blacks and two grays). That added up to twelve pups between the two sites.

42 led down to the road, and four of the pups at the main den followed her. It appeared she was trying to consolidate the family at the rendezvous site. She crossed the road and continued south, but the pups lingered on the road. 42 grabbed a stick and came back toward the pups. They ran to her, thinking they could steal the stick from her. That was at 6:46 a.m. She led them to the creek. With years of experience at getting pups across creeks and rivers, 42 knew how to trick the pups into following her.

It took her many attempts, but using a series of different sticks, 42 lured the pups across Soda Butte Creek and then through the wider Lamar River. When the pups balked at a crossing attempt, she would come back to them with a different stick and show it to them. Intrigued by the newer toy, the pups would run after her, and before realizing it, they were once again in the water and wading or swimming toward her. The four pups made it to the west side of the river at 9:07 a.m. It had taken the alpha female two hours and twenty-one minutes to accomplish that mission. Now ten of the twelve pups were at the rendezvous site.

Some of those pups later recrossed the river, and 42 patiently rounded them up and took them back to the rendezvous site. By the afternoon of July 6, there were seven black pups and two gray pups at the rendezvous site. The other three pups were up at the den forest. On the following evening, eight of the nine pups to the south followed 42 and 255 up the Chalcedony drainage. One gray female pup remained at the rendezvous site. In past years, I had seen other pups stay behind in that meadow and fend for themselves. Wolf pups can survive on their own for some time if there are enough voles and grasshoppers for them to hunt.

Six of the adults, including the alphas, returned to the valley on the tenth. They found a sickly bison cow west of the rendezvous site and killed her. 42 and U Black went to the gray pup staying at the rendezvous site and fed her. Half Black and 21 later also regurgitated to the pup. The pack was mindful of that pup and taking good care of the strong-willed little female.

302 returned to Lamar Valley two days later. He went up to the den forest that morning, then showed up at the rendezvous site in the evening. 302 spotted the lone gray pup and followed her up into the trees. The next morning, 302 was gone, but the pup was still there. That day Dan flew and found 21, 42, and the other collared Druids at the Opal Creek rendezvous site. He saw at least five pups but thought there were more in the trees.

On the seventeenth of July, 21 and three black adults returned to the valley and fed on the bison carcass. Later 21 went to the rendezvous site and gave the gray pup some meat. She had been there for a week at that point. Then he moved toward Chalcedony Creek, a route that would take them to the rest of the pack at upper Opal Creek. The pup followed him and I lost them in the trees. The other adults later took the same route. All the adults and pups in the family would soon be together up in their high-elevation summer hunting grounds. It was looking to be a good year for the Druids and their alpha pair.

17

Coexistence

NOW THAT THE wolves were out of the valley, several of us from the Wolf Project went out to the carcass of the bison the Druids had killed on July 10 and found that it was a yearling cow. The metatarsal bone in the lower part of a hind leg was broken, just above the hoof. It appeared to have been broken some time earlier, partly healed, then was broken again during the fight with the wolves. That leg injury made the cow vulnerable to wolf predation, in contrast to the vast majority of bison who are strong and healthy, too powerful for wolves to take down. Every edible bit of that carcass had been consumed.

A few days later, I got signals from 302 in Lamar Valley in the early morning, but did not spot him. The signals faded to the west. I figured he had gone to the Druid den and rendezvous site during the night, failed to find any wolves, and was now heading back to his family's territory. I drove west and spotted him a few miles west of Tower Junction.

The following morning, I got his signal from the Leopold den area.

The Druids came back to the Chalcedony Creek rendezvous site in late July. There were four adults (21, 42, 253, and 255) and eight pups: four grays and four blacks. 21 romped around with a black pup, then sparred with a gray pup. Later he was with three pups and pretended to run away from them. He ran slowly so they could keep up as they chased him. The other five adults soon arrived. With the first four adults and eight pups, that added up to seventeen wolves. The next day, a fifth black pup joined the group. That turned out to be the count of surviving pups: five blacks and four grays. Later 253 shared a bone with a pup. Both chewed on opposite ends of the bone at the same time.

I saw something new that day. Yearling 286 came back to the rendezvous site with a sandhill crane chick in her mouth. A small black pup and a much bigger pup ran toward her. The little black reached 286 first, grabbed the prize, and sped off with it. The big pup went after the other pup and tried to steal the bird, but the small black outran its sibling and kept the crane all to itself.

In August 302 continued his pattern of being with the Leopold wolves to the west, then traveling east to Lamar Valley. I got his signal in his family's territory on the second, then saw him alone at Chalcedony Creek on the fourth. The next morning, I had eight Druid adults and all nine pups at the rendezvous site, and I was still getting signals from 302 in the area. Wayne Kendall spotted him to the east, on Dead Puppy Hill. The Druids howled and 302 looked that way.

On the sixth, 302 was at the Chalcedony rendezvous site near the Druids. He watched them from a distance for an hour, then traveled back toward the Leopold territory. I thought about how he was continually risking his life by being close to the Druids. At any time, 21 and 253, along with other adults, could chase and catch him, then beat him up. He must have sensed that some of the nine Druid pups were his and had a paternal instinct to visit them.

I spoke with an instructor from the Yellowstone Institute and he told me about a backpacking trip he had taken recently. Near the Opal Creek rendezvous site, he had found a dead gray pup that had been partly consumed by a bear. He couldn't tell if the bear had killed the pup or found it already dead. It must have been one of the Druids' missing pups.

On August 20, I spotted the Slough Creek wolves in Lamar Valley near Amethyst Creek, just two miles from the Druids' Chalcedony rendezvous site. There were twelve wolves in the group: five adults and seven pups. The pack stayed there and used the area as a rendezvous site. A big bison died of natural causes on August 26 near there. That evening, three adults from the Slough pack fed on it. The pack continued to feed on the massive carcass for many days during a period of hot weather.

As far as we knew, wolves do not normally get sick from eating spoiled meat or animals that died of disease or infection. Dogs are classified as having acidic stomachs, and wolves living in the wild would likely have stomachs at least as acidic as dogs' stomachs. That allows meat to be digested quickly. The shorter digestion tract of wolves and dogs compared with humans allows for fast absorption of nutrients and moves waste material through the digestive tract at a rapid rate. The process minimizes the chances of bacteria in the meat infecting the wolf since it passes through so efficiently.

On the thirtieth, seven adult Druids were at the big bison carcass. 21 and 42 did a lot of scent marking, probably over spots marked by the Slough alpha pair. 253 stayed back at the Chalcedony rendezvous site with the nine pups. The Slough wolves were hanging out a few miles southwest of the carcass. They howled, but the Druids didn't seem to hear them. Later most of the Druids went back to their rendezvous site. The two yearlings, female 286 and the gray female, stayed at the site and continued to feed.

A black female from the Slough group approached the carcass without realizing the two Druids were there. She did see them when she came closer. Dropping into a crouch, the Slough female approached 286, who stood in a dominant posture. The incoming black lay down when she arrived, then rolled on her back under 286 to acknowledge her superior status. The gray yearling came over and also stood over her. The Slough wolf jumped up and gave 286 a submissive greeting. After that she climbed up on the huge bison carcass and fed. The interaction between the wolves indicated that the black Slough female was a former Druid and therefore a packmate to 286 and her sister. Later, three more Druids arrived, including 255 and Half Black, and they also were accepting of her. The Slough wolf fed and later went back to her pack and regurgitated to one of the pups.

On the first day of September, five adult Slough wolves and four pups were west of Amethyst Creek. All nine Druid adults were on the bison carcass. The adult Sloughs traveled toward the carcass with the four pups following them. When they got closer, the Slough adults, led by alpha female 217, ran toward the site, anticipating a meal. The pups followed.

One of the Druids must have seen the Sloughs, for it charged at the approaching wolves. The other Druids followed its lead. The five Slough adults responded by running straight at them. 217 veered off just before reaching her birth family. The other four adults in her group continued racing forward and ran into the middle of the nine Druids. They scattered when they realized how outnumbered they were. Wolves ran back and forth, and two of them got into a fight. 21 stayed close to 42 during the conflict, seeming to guard her.

In the middle of the chaos, the Druids came together for a rally. Then they spotted 217 and ran all out after her. They could have caught and killed her, but instead they pulled up and let her go. Then they pursued alpha male 261, but allowed him to escape as well. 21 targeted an adult gray male and went after him, then broke off the chase. It seemed 21 and his family were satisfied when they saw the other wolves running away and did not pursue them further.

During the interaction, the pack's nine pups remained at the rendezvous site. The Slough black female who had met up with 286 and seemed to be a former Druid wandered into that area and sniffed around. The pups ran toward her and the female raced away from them. The pups probably thought this was a Druid adult bringing them food from the carcass. When the black stopped and faced the pups, they surrounded her and wagged their tails. She walked off and they followed. The female stopped again, and they clustered around her. One of the pups tried to lick her face, hoping for a regurgitation. The black Druid male yearling trotted into the rendezvous site and the pups ran to him. He gave them a regurgitation. The Slough female went to him, and the two related wolves had a friendly meeting. Pups then ran over and licked her face. Later she went to the carcass and met up with the other Druids. She rolled on the ground under them, then got up and romped around as the Druids wagged their tails at her.

For now the Druids seemed to have an understanding with the Slough wolves, similar to the one they apparently had with the Agates. Each of those other packs was led by a former Druid female, so all three packs were related. I also sensed that the three packs' alpha males—21, 113, and 261—felt secure in their positions and had good judgment about when it was necessary to fight rival wolves and when it was not.

That evening the Druid adults brought their nine pups to the bison carcass, but the pups seemed hesitant to get near it. This was probably the first time they had been at such a massive carcass. They would cautiously approach the site, sniff it, then back off. Some nipped at one of the legs, like they were testing to see if the animal would jump up and attack them. By the next morning, the pups had overcome their fear and were happily feeding.

302 was back in Lamar on September 5. He must have gotten the scent of the bison carcass, for he went right to it. Druid female yearling 286 was approaching the site and he ran off when he saw her. She was black so he might have mistaken her for 21 or 253. I had observed them together back

in July, so they had previously met. 302 soon came back and 286 playfully jumped around him as he wagged his tail at her. Then he sniffed her face and licked her mouth. After that she repeatedly jumped up on his back. They ran off together and I lost them in some trees. 302 still had whatever it was that made females flock to him.

The next morning, I saw 302 in Lamar traveling with two of the females he had mated with in the spring—255 and Half Black—and a new companion, the gray female yearling. 255 led the group of four wolves toward the rendezvous site. When they arrived, no other Druids were in sight. 302 stopped, looked around, then ran off. A few moments later, I saw 21 coming out of the trees. He was stalking 302. I followed 302 as he traveled farther away, then bedded down and watched the Druids from a distance. Later he got up and headed home to his parents' territory.

302 came back to Lamar seven days later and had a friendly meeting with 255. The following morning, he was halfway back to Blacktail Plateau. That had become his set pattern: walk the twenty-five miles to Lamar, visit with some of the Druid females and check on pups, then travel back home. He always avoided 21 and 253. I had to give him credit for how much effort he put into maintaining his relationships with the females and pups.

On October 10, Carol Rickman called to tell me that she had wolves south of Slough Creek. I joined her and saw a gray female with two black pups. They were looking intently to the southeast. Soon I spotted 302 and a black female approaching from that direction. They joined the gray female and the pups. 302 seemed to be courting the black female. Later it appeared that the gray drove the black off, and 302 ended up with her and the two pups. They all went to the old bison carcass in Lamar and scavenged on it. Both pups were very comfortable with 302. They probably already knew him. The gray female flirted with 302.

A few days later, 302 and his newly formed group were in the Elk Creek area, a mile west of Tower Junction. Female 251, who had been born in the Druid pack and later spent time with the Agate pack, had denned nearby and we had seen her there with five black pups. She died in late June and apparently this gray female had been taking care of the two pups that survived their mother's death. Since the gray was uncollared, I didn't know which pack she was from. I had guessed that 302 knew young females in packs other than the Druids, and this unknown female seemed to prove that. I wondered how many females he had romanced during the two mating seasons he had lived through as an adult. Since he seemed to have a short attention span, it was probably a lot.

In late October, the Druids chased two big bull elk. The bulls stopped and faced the wolves. As the Druids approached, the nearest bull lowered his head and lunged with his antlers at the closest wolf. He repeatedly charged at the wolves and tried to gore them, but they managed to dodge all those thrusts. I had examined a dead Druid wolf several years earlier with an antler wound in his chest and knew how deadly those lunges could be.

In my early years in Yellowstone, I heard of a man taking photos who had gotten too close to a big bull elk. It was the mating season and that bull was in an aggressive mood. He charged at the man, lowered his head, and lunged forward with his huge antler rack. The man had backed up against a wooden fence and couldn't dodge the thrust. The spread of the antlers was so wide that the points on each side ended up jamming into the fence just barely to the man's right and left. The story ended there and I never heard how the guy got away while the bull struggled to get his antlers unstuck from the fence.

Unlike the horns on bison and bighorn sheep, antlers are shed in late fall and regrown in early spring. Shed antlers can be sold for decorative purposes, and the Yellowstone rangers frequently catch people trying to sneak out of the park with a load of stolen antlers. One day I was searching for wolves from a ridge high above Lamar Valley and saw a man picking up antlers and hiding them in thick brush close to the road. He likely intended to put them in his vehicle when no one was looking. I used my park radio to report the situation to the local ranger and he caught the guy. As the ranger gave him a ticket, the man asked how they knew what he was doing. Protecting his source, which was me, the ranger looked up in the sky, pointed at a random spot, and said, "See that satellite up there?" The guy looked that way and nodded his head.

302 was still with the gray female and the two black pups on October 28, a period of eighteen days. For 302, that qualified as a long-term relationship.

On the last day of the month, I found the Druids back in Lamar Valley. 253 had a bad limp on his left front foot. It was his right front paw that had been caught in the steel trap in Utah, so this was a new injury. Despite that hindrance, he was leading the pack the next morning. He mostly held his left paw off the ground as he traveled. When he rested, he licked it. This was his third injured leg.

That day I got a report that the Druids had dug up a carcass that probably had been buried by a mountain lion, for they customarily cover kills with dirt. The wolves pulled it out into the open and frequently looked up into a nearby tree as they fed. They must have been seeing the lion there, waiting for them to leave.

On November 9, the Druids went back to the bison that had died on August 26 and scavenged on the ten-week-old carcass. No meat was left at the site so they mostly chewed on the bones. When they left nine hours later, 253 kept up with the other wolves but held his left front paw off the ground as he walked. It took until nearly the end of December for the paw to heal to the point that he was putting weight on it most of the time.

21 and 42 did some flirting in the third week of November. He put his chin over her back, then she rolled on the ground and playfully kicked up at him. When she got up, 42 wagged her tail at him. Both were now eight and a half years old, about sixty-eight in human years, and still affectionate with each other.

Doug did some radio collaring that day. He went after the Geode wolves and darted three of them. One was a black adult female. When Doug got out of the helicopter to process her, he found her hiding under a boulder. It turned out that the dart didn't go into her, so Doug jabbed her with a new dart by hand, then collared her when the drugs took effect.

I saw 302 with the gray female and the two pups in the Elk Creek area on December 8. He had now been with her for nearly two months. The Agate wolves were nearby. Doug did a flight the next morning and radioed down to me that he had both groups close to each other in that area. I saw a fresh elk carcass there, south of the road. Doug added that he had not gotten 103's signal or seen her for several months. I later checked my records and found that I had last seen her with the Agates on October 5. I spotted 103 away from the pack three weeks later, and the last time I got signals from her was on November 11. She would be over six and a half years old if she was still alive. Without her, the Agate pack now had ten members.

The Wolf Project's early-winter study had started. After I left the Elk Creek area on December 11, Tim Hudson and Lisa Turner, who were observing the Geodes, saw 302's group go to the elk carcass. The four wolves fed for forty-five minutes, but seemed nervous, probably because they were getting the scent of the Agate wolves in the area. 302 and the female left, but the two pups continued to feed. The pups suddenly ran off, and the crew saw nine Agates chasing them. As the wolves were going out of sight, the pups were only ten yards ahead of the pursing wolves. Ten minutes later, the Agates came back into view and had a group howl.

The crew then heard bark howling from the north side of the road. That sound is made by a wolf in distress. The gray female came into sight. She had probably sensed that the pups were in trouble. 302 wasn't with her, and the Geode crew couldn't get his signal. Later they found the tracks of a single wolf in fresh snow on the road, a half mile to the west.

The tracks showed that the wolf was running away from where the Agates would be. Since the gray female was visible to the north, those tracks must have been made by 302.

Two days later, Dan Stahler called me. He had done a flight that day and seen 302 and the gray female about five miles west of that area. No pups were with them. The Geode crew discovered that the Agates had killed one of them and we never saw the second pup, so it likely died as well. That meant that none of 251's pups survived.

Dan was finishing up a PhD on wolf genetics at the University of California, Los Angeles. His team analyzed a DNA sample taken from the dead pup and found that his parents were 251 and 302's younger brother, 301. I never saw those roving Leopold brothers at 251's den site or got their signals there, so they likely had done either nothing or very little to help her with her litter before she died.

I had to conclude that when the Agates chased the pups, 302, the acting alpha male in his small group and uncle of the pups, had saved himself by running away. The most basic responsibility of a male wolf is to defend his pack, and 302 had repeatedly failed in that duty. For some time, I had thought of comparing the relationship between 21 and 302 to the one between Batman and the Joker, but then I came up with a better analogy. In *Superman* comics, there is a long-running character called Bizarro who was an attempted copy of the superhero. He had all of Superman's powers but always acted in the exact opposite way Superman would. That was 302: a wolf who was the opposite of 21.

When I talked to park visitors about 302, I found I had to choose my words carefully. People loved the bad-boy aspect of his character and if I said anything negative about his behavior, they were quick to come to his defense, saying things like "But he didn't mean it" or "But he was trying the best he could."

Around that time, there was a change in leadership in the Slough Creek pack. 217, the founding alpha female, was now subordinate to an uncollared black female. We thought the black, like 217, was a Druid.

On December 15, Doug Dance, a Canadian wildlife photographer, spotted 302 and his gray companion south of Slough Creek. Half Black was nearby. 302 went to her and they had a very friendly meeting. The gray female was hesitant about this new rival and kept her distance. The rekindled relationship between 302 and Half Black was interrupted when 21 came on the scene and chased off 302.

But as I had seen in the past, 21 could not control his daughters. Soon Half Black was back with 302, and her sister U Black was with them. The gray female didn't join in when the trio played together. U Black later returned to the Druids, as she had done the year before, but Half Black stayed with 302. It appeared that he had moved on from the gray female, and we didn't see her with 302 after that. The next day, 255 reunited with 302. He ran away when 21 charged at him yet again.

The following morning, when the Druid adults were away from the Chalcedony rendezvous site, 302 visited the five Druid pups there and wagged his tail at them. The pups licked his face and jumped up on him. The main group of Druid adults howled and 302 and those pups howled back. The pups moved off and 302 followed them. When they got tired, he bedded down with them. This suggested that he had some sense, perhaps a paternal one, that he should help care for them.

Later, when the main group of Druids arrived at the rendezvous site and met up with the pups, 302 ran away. I saw him off in the distance, sitting up and watching. 253 charged at him with the other Druids just behind him, and 302 raced off. It was a poignant moment to witness. 302 seemed to be trying to establish a connection with those pups, but his past behavior understandably caused 21 and 253 to drive him away.

In the following days, we frequently saw 302 with 21's three daughters: 255, U Black, and Half Black. When he wasn't with them, he often was near the Druids and howled a lot, seemingly to contact those females. He also regularly visited the pups. When a gray pup met up with 302, it licked his face and excitedly jumped up and down in front of him. Later the pup did a play bow to 302, hoping that would start a play session. But 302 was more interested in the nearby U Black, and the two adults did a lot of playing together. 21 and 253 repeatedly tried to catch him and beat him up, but 302 always managed to slip away. On the last day of the year, I got 302's signal back in the Blacktail Plateau area. He was taking a temporary break from 21's daughters.

I added up the total number of wolves in the Druid pack and the four new adjacent packs that were headed by alpha females who were former Druids. It came to fifty-two wolves.

At the end of 2003, the Wolf Project estimated that there were 174 wolves in fourteen packs throughout the park. That turned out to be highest population ever for Yellowstone.

The original wolf population, before wolves were eliminated from the park, was around one hundred. The restored population reached that level in 2000, then gradually rose to its peak density in 2003. After that the wolf numbers trended down and by 2009 tended to hover near one hundred. The counts above a hundred took place when the park's elk numbers were above the park's carrying capacity. When the elk population went down to a more sustainable level, wolf numbers reverted to the long-term average.

PART V

2004

Yellowstone Wolf Pack Territories
2004

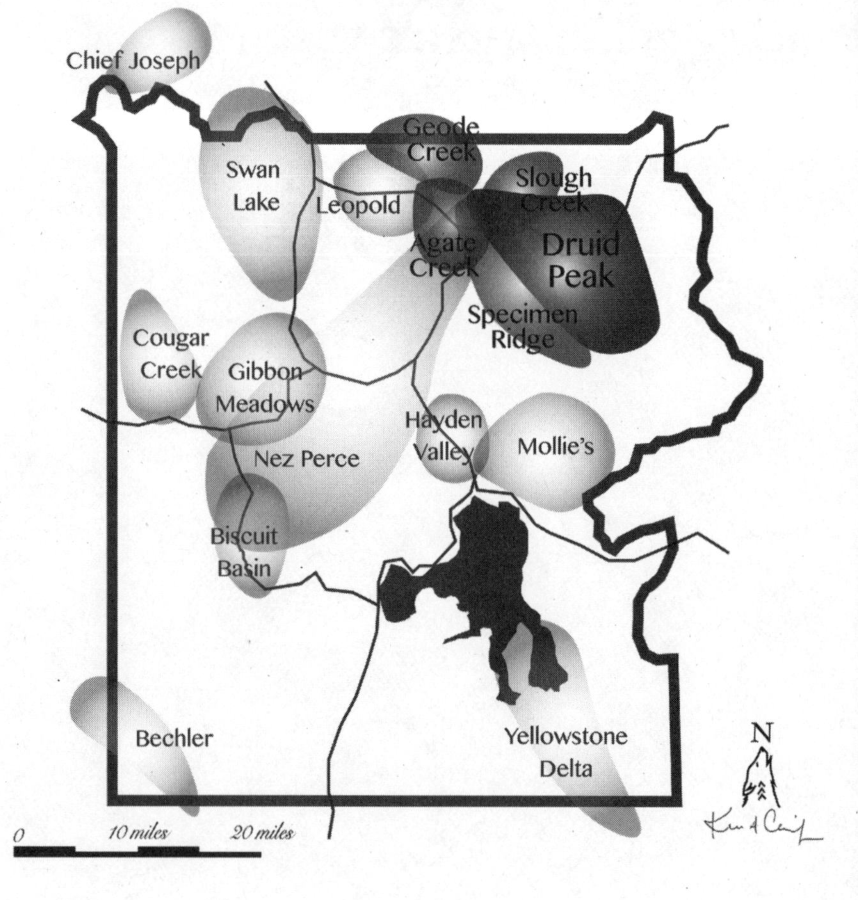

PACKS INCREASE AND decrease in size over the course of a calendar year. These charts show the main pack members in any given year. M=male and F=female. An asterisk (*) indicates a female thought to have denned. The pack of origin for wolves joining from other packs is indicated in parentheses the first time a wolf is introduced. Squares indicate adults and yearlings. Circles indicate pups.

Druid Peak Pack

255F mated with three roving males and left the pack to have her pups. When none survived, she returned to become a Druid once again.

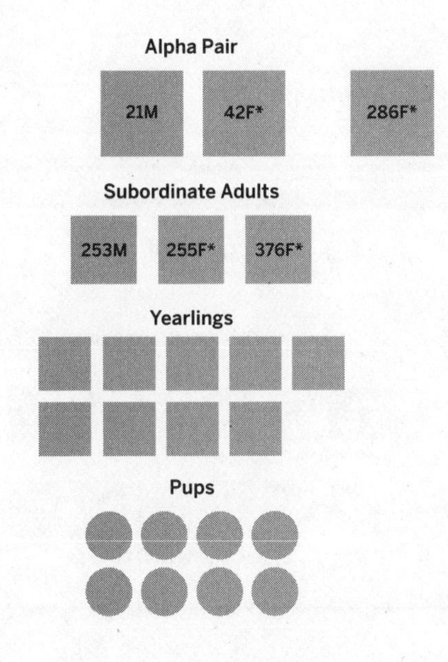

Specimen Ridge Pack, Formed 2004

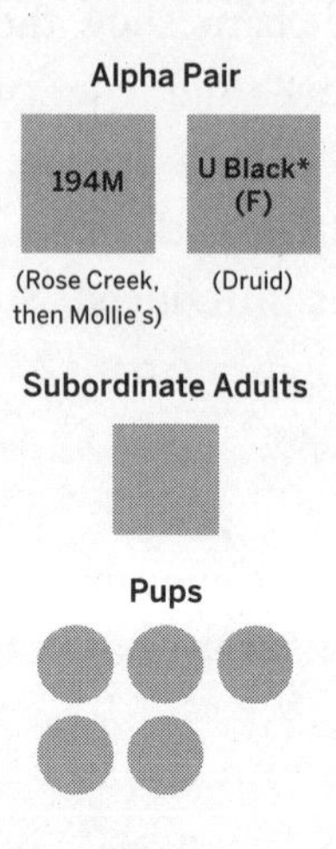

Slough Creek Pack

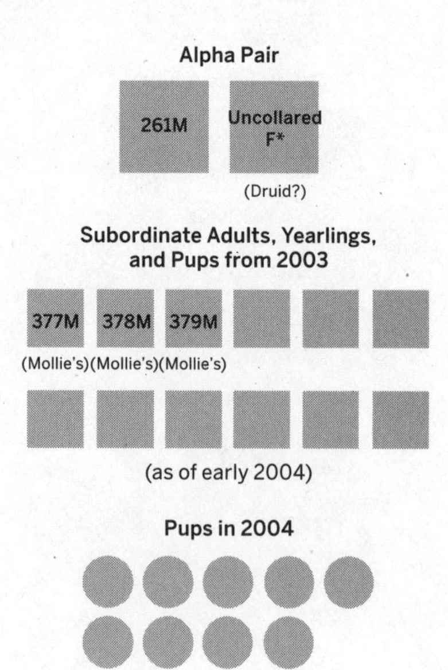

Unattached Males

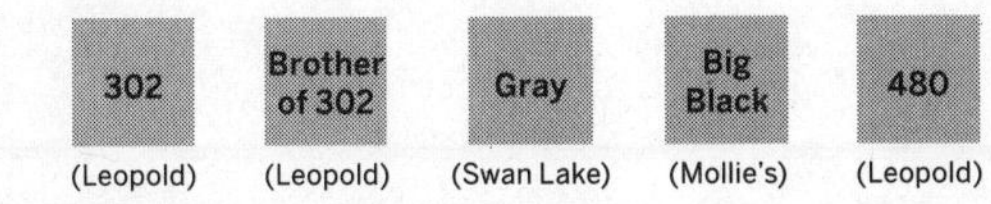

Female Wolves Seen With 302

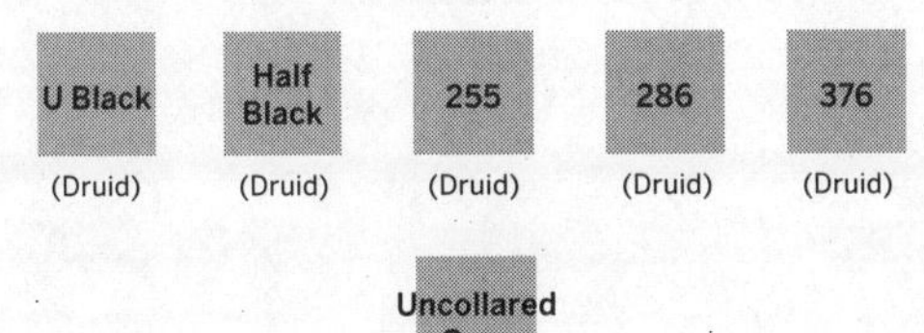

18

January

IN JANUARY 2004, the Druids still numbered seventeen, including nine pups. Six pups were captured and radio-collared. DNA analysis found that 302 had sired five of them. One of the five was born to 302 and 255. The four others were born to 302 and either U Black or Half Black. Neither female had been collared, so we didn't have any genetic material from them. 21 and 42 were the parents of the sixth pup. That meant that at least five of the nine surviving 2003 Druid pups were fathered by 302, and most of the pups 21 had raised that year had been sired by 302. I had seen 302 hang out with the Druid pups at the rendezvous site and wondered if he had a paternal instinct to visit them. Now we knew that he was the father of some of them. The gray female yearling was also collared and became known as 376.

On the first day of the new year, I saw U Black flirting with 302 on the north side of Little America. The main

Druid group was nearby and they howled. 302 howled back. I swung my scope back to the Druids and saw that 21 was staring directly at 302. It was not a friendly stare.

The next day, there was an uncollared black male with 302 and U Black. He looked younger than 302 and acted subordinate to him, so he was likely one of 302's brothers. Soon after that, U Black returned to the Druids as she had done the previous year. It seemed that she didn't see 302 as suitable material for a permanent mate. Her short-term associations with him contrasted with the multiyear relationship of the Druid alpha pair. 42 still liked to be close to 21. One morning after traveling for a while, 21 bedded down and 42 went right over and lay down beside him.

Doug did a tracking flight on January 11 and got a mortality signal from the deposed Slough Creek alpha female and former Druid, 217. Nine other Slough wolves were in the area and they had a fresh elk carcass. The wolves and the kill site were on top of a cliff, and 217's body was below them. A crew later went to that site and found wounds on her that indicated she had been killed by other wolves. Perhaps 217 and the new alpha female had gotten into a fight that resulted in 217's death.

By January 20, two more dispersing males arrived in Lamar Valley. One was 194, a big black from the Mollie's pack. He had been born to 8 and 18 in the Rose Creek pack in 1997, so 21 would have helped raise him. When he was older, he left his family and joined the Mollie's. I later found out that 261, the alpha male of the adjacent Slough Creek pack, was one of 194's sons. A gray male accompanied 194, most likely another Mollie's male. U Black was with them too.

That meant that she had been with four males that month: 302, 302's younger brother, 194, and the other Mollie's male.

I saw U Black's sisters, 255 and Half Black, with 302 and his brother the next morning. Younger female 286 joined them. 302 tried to mount Half Black, then did the same with 286. She flirted with both Leopold males. It looked likely that all four of those Druid females were going to get pregnant by outsider males. I wondered which ones would start their own packs and which ones would come home to their family to have their pups.

On January 22, I got a report of a mating tie between 21 and 42. The next day, yet another dispersing wolf showed up in Lamar Valley, a gray male from the Swan Lake pack, the group that lived near Mammoth Hot Springs. 255 and yearling 376 ran over and playfully romped around him. Then Half Black joined the group and the male had three females flirting with him. 21 saw what was happening and charged at him. 21 had to chase the latest of his daughters' suitors three times before he finally moved off.

I saw a sixth outsider male in Druid territory a few days after that. It was a big black uncollared wolf. We didn't know where he had come from. 255 and Half Black were already interacting with the new arrival. Later the Swan Lake male approached the group and the big black chased him away. The next day, 21 and other Druids drove the big black away from his daughters, but Half Black went back to him once her father left. There was another tie between 21 and 42 on January 28.

One day late that month, I was with several park visitors watching the Druids. One woman from Ireland knew a lot about Celtic family names. Both of my parents are descended from Scottish people. On my father's side, the MacIntyre/

McIntyre clan lived in the Highlands. The Irishwoman explained that in some Celtic languages, my last name could be translated as "Son of the Land." That phrase was also a poetic way of referring to a wolf in ancient Scotland and Ireland. All that was unknown to me, but now I realized that our clan name predicted my destiny.

On the last day of January, I saw two gray males with 255 and Half Black. Both had come up from the Mollie's pack, possibly to join their brother, Slough alpha male 261. They were 377 and 378, the seventh and eight dispersing males to court 21's five adult daughters that winter. 378 was later seen mating with Half Black.

After that, I drove to Slough Creek and saw eleven of the Druids. 21 and 42 were flirting a short distance from the other wolves. She put her chin over his back, as I had seen her do in Bob Landis's film of the pair's first meeting in fall 1997. He sniffed her rear end when she averted her tail. At 10:47 a.m. they got into a breeding tie. As she settled into a bedded position, he was draped over her back with his two front legs positioned so it looked like he was hugging her. The pair broke their tie at the twenty-one-minute mark. 21 then affectionately licked 42's face. She lay down and he bedded down next to her.

The two wolves had been together every day for over six years. They were just under nine years old, so they had spent about 70 percent of their lives with each other. 21's coat had turned half gray and his face was entirely gray. 42's black fur was now the color of gray flannel underwear. They were like some married couples who had been together for many decades, who had grown old together, but were still lovers and best friends.

That reminded me of someone. Years earlier when I was working in Joshua Tree National Park in the California desert, there was a day when I was staffing a small visitor center. I was the only one in the building. A car pulled into the lot and a man got out and entered the building. He looked familiar and after hesitating a bit, I asked, "Excuse me, but are you Johnny Cash?" He turned to me and politely said he was.

We talked briefly, but I sensed he was in a sad mood, so I left him alone. He soon walked out and drove off. I knew that he had been married to June Carter for a long time and that they were devoted to each other. She wasn't with him that day and perhaps that accounted for his subdued demeanor. He wasn't a famous star when he spoke with me, just a lonely man apart from his wife and soul mate. A few years later, I heard that she had passed away. Johnny never recovered from her loss and died four months later. That seems to happen often with longtime married couples who are especially attached to each other: one dies and the surviving spouse passes away soon afterward. There is a word in Okinawa, Japan, that aptly describes individuals in that situation: *ikigai*. It can be translated as "a reason for being" or "the reason you get up in the morning." If you lose your lifelong partner, there may not be a reason to keep on living.

Later in the day, I returned to where I had seen 21 and 42 mating. The Druids were bedded down just west of Slough Creek. Everything was peaceful when I left the pack at dark and headed home. That night something happened that changed everything.

19

February

EARLY ON THE morning of February 1, I got a report that 21 had been seen to the north of the Lamar River in Little America. He was alone, which was unusual. I then heard that people at Slough Creek were seeing other Druids there. I went up on Dave's Hill and saw Druid wolves bedded down across the creek. I got signals from 253 and two of the pups. 21's collar was no longer working so I couldn't get a signal from him. I should have been getting 42 but couldn't pick her up. Carol Rickman called to report that she had just seen seven wolves on the top of Specimen Ridge. I got signals from the Mollie's alphas from that direction, so it had to be their pack.

I later went to the middle of Little America and saw that twelve Druids had joined 21. They were at a fresh bison calf carcass. The wolves howled off and on. I was getting loud Druid signals, but still nothing from 42. I looked south, toward Specimen Ridge, and saw some of the Mollie's wolves.

Then I spotted the Mollie's alphas on the west end of that ridge. The alpha male was looking toward the howling Druids. I got a report that one of the Mollie's wolves had blood on his muzzle and a wound on his side. That made me think there might have been a fight between the Druids and the Mollie's during the night. The two packs stayed in their respective locations for the rest of the day. I left as it was getting dark and tried to get a signal from 42 one last time but failed.

The next morning, I found twelve Druids at Slough Creek. 21 was there but 42 was still missing. Two Mollie's wolves were up on Specimen Ridge. I drove to the Tower Ranger Station and called the Wolf Project office. Deb Guernsey told me that Dan Stahler had just left the airport on a tracking flight. I asked her to get a message to Dan to call me on the radio.

I drove to a spot where I had a good view of Specimen Ridge. I heard seven Mollie's had been seen there. I tuned to 42's radio-collar frequency and got her toward Specimen Ridge, but it was a mortality signal. That meant the electronics in her collar had not detected any movement for four hours and transmitted a faster-than-normal signal. There have been times when a collar malfunctions and falsely transmits that signal. I hoped that was the case now.

I heard the tracking plane. Dan called down to tell me that he had spotted the Mollie's wolves in Little America and thirteen Druids at Slough Creek. I saw him circling the west end of Specimen Ridge, near where I had seen the Mollie's alphas the day before. When he radioed me, Dan said the words I did not want to hear. He had seen 42 and confirmed that she was dead. It looked like she had been killed by wolves.

Dan later told me there were signs that a big struggle had taken place where he had spotted 42, meaning she had put up a good fight to save herself. A few days later, Doug Smith and several other Wolf Project staff skied and snowshoed up there. Two black wolves that looked like Mollie's were in the area. Doug collected 42's skull for research purposes and left the rest of her remains.

Matt Metz had come out to help. He called from Little America to say 21 was howling near the base of Specimen Ridge. I figured he was calling for 42, not knowing she was dead. The other Druids were north of the road and they joined in. I drove to Little America and saw 21 running across the road to join the rest of his family. On the way, he stopped and howled. Then I spotted seven Mollie's wolves at the base of Specimen Ridge.

I watched the Druids come together. They had a group howl and rally that centered on greeting 21. Female yearling 286 averted her tail to 21. Her father was a Nez Perce wolf, so he could have bred her, but after sniffing her, he walked off. I guessed 21 was concentrating on finding 42 and not thinking of mating.

I could see the Mollie's wolves having a rally in the distance. When they finished howling, they stared in the direction of the Druids, then traveled to where 21 had been earlier. I got a report from park volunteer Bob Weselmann that U Black was in that area. She howled back at the Mollie's. Later she met up with a big gray from that group and flirted with him but ran off when other Mollie's wolves came toward them.

As I drove home at the end of the day, I tried to figure out what had happened. I had seen the Druids at Slough Creek late in the day on the thirty-first and everything had seemed normal. They were near that area early on the first, but I didn't get 42's signal and 21 was seen alone, to the west of his pack. We saw the Mollie's wolves on Specimen Ridge and later learned that 42's body had been found south of their position.

I imagined that the Druids and Mollie's got into a fight during the night, probably at Slough Creek. As 21 fought with some of the big males, other Mollie's wolves went after 42. She probably ran west and may have been attacked during the chase. I think she then went south across the Lamar River, crossed the road, and continued up Specimen Ridge to the site where Dan spotted her, and the other wolves caught up with her there. The sighting of 21 the following morning, to the west of Slough Creek, away from the other Druids, suggested that he followed her scent trail to that area, but probably lost it when she crossed the river. He went back to his pack and later returned to that area with the other wolves. They did a lot of howling from there, probably trying to contact 42. When I saw the Mollie's that morning on top of Specimen Ridge, they probably were coming back from their fatal attack on 42.

The Druids and Mollie's wolves had been feuding since the spring of 1996, nearly eight years. Back then the original version of the Mollie's, known as the Crystal Creek pack, had claimed Lamar Valley as their territory. The Druid wolves, probably led by their aggressive alpha female, wolf 40, had attacked the Crystal Creek pack at their den and killed their alpha male and the litter of pups. They had also injured the alpha female, but she survived. She and a young male abandoned their territory to the Druids and established a new one in Pelican Valley. Several generations later, it appeared that the current members of the Mollie's pack were continuing that feud by killing 42, the last of the Druids who had been alive back then. The Druids had killed one of their alphas, and now they had killed a Druid alpha. But knowing 42 so well, I thought it unlikely that she had participated in the fatal attack on the Crystal Creek's alpha male.

I saw twelve Druids back near their main den early on February 3. I wondered if 21 had led them back there in hopes of finding 42. Late that day, 21 took the pack west. The next morning, they were south of Slough Creek, likely looking for her there. 286, now the dominant female in the group, was at the peak of her cycle. 21 tied with her in the late afternoon. Over the next few days, the two wolves mated three more times. Since 21 was the alpha male, he was also the breeding male. If he didn't breed her, an outsider like 302 would, and once again 21 would be raising another male's pups.

During that time, we frequently saw U Black with Mollie's male 194 and the gray male we thought was his brother. 255 was consistently with an uncollared black Leopold wolf, the one we thought was 302's younger brother. That pair mated on February 7. We didn't see 302 and I thought that the younger Leopold wolf, who had become dominant to him, had probably driven him away. There was too much competition in the area over the Druid females, and 302 likely was off looking for other females that he could have to himself. 302 didn't seem to have what it took to challenge other males and win.

Later 302 found a gray female from an unknown pack and hung out with her. A younger black male, probably another Leopold wolf, joined them and was subordinate to 302. I wondered if this was the same gray female 302 had been with in late 2003, the one who took care of 251's pups.

In the second week of February, 253 left the Druids. We found him at the west end of Lamar Valley traveling west, probably looking for a mate. As he was about to leave the valley, I saw him stop and howl. The Druids, a few miles to the east, howled back. 253 then continued his journey west. That left only two adults in the Druid pack: 21 and 286. All the other wolves were pups.

255 was no longer with 302's younger brother, but with the big black uncollared wolf she had met up with earlier. He seemed to be from the Mollie's pack. The two mated on February 9. I later saw that 302's brother had wounds on his hip and shoulder that must have been inflicted by the larger male. Druid female 376 temporarily traveled with 302's brother, but she soon left to go back to her family and he was alone again.

255 and the big black were joined by two other Mollie's males, 378 and 379. Four days after 255 tied with the big black, she was seen mating with 379. That was the third male she had bred with that month.

I got 253's signal in the Leopold pack's territory on February 10. There were plenty of young females in that pack so we hoped that he would find one to pair off with. That day there was a tie between U Black and Mollie's male 194. During that time, I often got the Druids' signals at their den and continued to wonder if 21 went there in hopes that 42 would return to that site.

On the fifteenth, 253's signal was in the Mammoth Hot Springs area. There was another signal indicating that a Leopold female was also in the area, and the two could have been together. But the next morning, I found him back in Lamar Valley, alone. Roger Stradley flew over and called down to tell me that he saw thirteen Druids south of 253's position. Early the next morning, I got 253's signal from the Druid den forest, and on February 18, I saw him back with the Druids. This was the second time he had attempted to disperse only to return to his pack.

302's younger brother continued to have a run of bad luck. I saw him on Dead Puppy Hill on February 19, and Druid female 376 was with him once more. The big black from the Mollie's who had taken 255 away from him arrived, and I saw the two males fighting over 376. The larger male pinned the Leopold wolf and bit him. The pinned wolf jumped up and ran off. The big black went over to 376 and two minutes later was in a mating tie with her. After they broke apart, they both looked north. The male lowered his tail and quickly moved away. I saw 21, 253, and 286 charging in. They chased the black, then went to 376, who had been away from the family for some time. The older wolves sniffed around her and must have been getting the scent of the outsider male.

That big black male stuck around and his persistence paid off. In the afternoon, I saw that 376 had reunited with him. Once again 21 ran in and the big black fled. This time, as 21 was about to reach him, the outsider turned around and bit 21, then ran away with his tail tucked between his legs before 21 could attack him. 21 chased him and the rival male got away. Thinking the incident was over, 21 crossed the road to the north and went up into the den forest. But the black male turned around, found 376, and got in another tie with her. The next day, 376 was back with the Druids.

In the last few days of February, 255 was back with the big black male. U Black was with Mollie's male 194 and his gray brother. 302 and his gray female were also still together. Half Black had not been seen since the last day of January when she mated with Mollie's male 378. We never did see her again. With the return of 253 and 376, the main Druid pack now usually numbered thirteen: four adults (21, 253, 286, and 376) and the nine surviving pups from 2003.

42 had been gone for a month now. Some wolf watchers felt 21 was spending more time alone, away from the pack, and speculated he was depressed, like a dog who has lost a longtime canine companion or a human friend. I had the impression he was doing more howling than usual. I couldn't tell if 21 now realized she was dead or felt that she was just missing. If he thought she was missing, did he wonder whether she was wounded somewhere out there, trying to hang on while waiting for him to find her?

As I considered those possibilities, a troubling thought came to me. For over six years he had been with her every day and had protected her. She was his priority. Did he now feel that he had let her down? What good was it to be the pack's alpha male if he couldn't protect his mate? The psychological effect of thinking that he had failed her would have been devasting to a wolf like 21.

20

March and April

ON THE FIRST day of March, 302 and his gray female showed up in Lamar Valley. The Swan Lake male who had approached 21's daughters in January was south of the Druid den area, headed toward one of the Druids' gray female pups. She romped to him and they had a friendly meeting. Then 302 and the gray female arrived, and the Swan Lake male tucked his tail between his legs. 302 charged and chased him across the road. Another observer saw 302 attack and pin him. That surprised me because 302 usually ran away from rival males.

After that I saw 302 return to the gray female, and one of the collared Druid pups romped playfully around him. Genetic analysis determined that 302 had sired her. I wondered if 302 could tell from her scent that he was her father. If so, it made his uncharacteristic attack on the Swan Lake male more intriguing, for he would have been driving away a male wolf from his young daughter. That was exceedingly ironic, considering how many times 21 had chased 302 from his daughters. I then thought about 21. When he sniffed the pup, could he tell she had not been fathered by him, but by 302? Whatever he discerned from her scent, 21 seemed to treat her just like all the other pups.

The next morning, the Druids were at Crystal Creek. 21's daughter 255 had been alternating her time between traveling alone and being with the big black male from the Mollie's. That day I spotted her near the Druids, but when the pack started traveling, she moved off in a different direction. Eventually she split from the Mollie's male, but her bulging sides revealed their liaison had resulted in a pregnancy.

I paid special attention to how 21 interacted with the Druids' new alpha female, 286. One day when 21 was bedded down, 286 moved toward him. He got up and walked off without greeting her. I wondered if he had bred her as a duty so that the Druids would have a new litter of pups that spring. His relationship to her was nothing like the intimate one he had had with 42.

One morning in mid-March, 253 walked over to 21, greeted him, and licked his face. Then 253 held up his left front paw and 21 licked his son's foot. This was the paw that he had injured in November, four months earlier. It must have been hurting him that day. When the licking ended, 253 moved the paw closer to 21's face. That got him to continue the licking. The next time 21 stopped, 253 pawed at his father's face and 21 resumed licking the foot. After that the younger male held up his right front paw, the one that had been caught in the coyote trap, and 21 licked the area just above the paw, the spot where the jaws of the trap had closed in on him. When that was done, 21 licked 253's face.

I remembered when 21 had licked the face of another young Druid male, wolf 224, after he had been picked on by some of his siblings. That brought another incident to mind, the time when 21 went to a pup in poor health and hung out with him. 21 seemed to sense when others in his family were hurting, either physically or emotionally, and would go to them and give them special attention.

253 was now nearly four years old, about thirty-five in human years, while 21 would soon be nine. 21's daughters 286 and 376 were close to two years, and all the other Druids were about eleven months old. I got the sense that 21 felt most comfortable being with 253 and had a special bond with him. Perhaps that was partly because 253 was the wolf closest in age to 21. Another issue was loyalty. 253 had the same type of loyalty to his family that I had seen in 21 back when he was a young male helping his mother and 8 feed and protect their litter of pups in Little America. The father and son were a lot like each other.

I got the Druids' signals from their den forest on March 24, then spotted them sniffing around an opening among the roots of a dead tree. The wolves seemed hesitant about getting too close to that spot. We later found out that a female black bear had chosen that site for a den and had given birth to two cubs in it. The cubs and mother were still in the den and she was probably growling at the wolves.

That morning I saw 302 south of the Druid den area. He was running north, toward the road, and 21 was right behind him, about to catch him. Reaching the road, 302 ran down it to the west. He veered around two large vans and continued west. 21 ran up on the road and started to head that way but turned back when he saw the vans. Once again, he let 302 go.

Soon after that, 302 ended up with some of the Druid pups from the 2003 litters. They greeted 302 and jumped up on his back. A gray pup suddenly got aggressive with him. 302 nipped him in response, then ran off with his tail tucked between his legs. It was a startling sight to see a big adult male running away from such a small low-ranking pack member. 302's fearful reaction to that pup's aggression didn't seem to have any detrimental effects on how female wolves felt about him though. 376 ran after 302 and flirted when she caught up with him. Those pups would be a year old in a few weeks, so from now on, I will call them yearlings to distinguish them from the 2004 pups.

Our pilot, Roger Stradley, spotted U Black with Mollie's male 194 and the gray male from the Mollie's on a spur ridge on the north side of Specimen Ridge, just west of Crystal Creek. I saw the trio there again on the twenty-sixth. A few days later, we saw that U Black was denning on that spur ridge. There was a small opening in the forest, and we often saw her there with Mollie's male 194 and his gray brother. The trio were named the Specimen Ridge pack.

At the end of the month, 103's body was found near the road just west of Tower Junction. She had been hit by a car and killed. 103 was nearly seven years old and had lived well past the average life-span for a Yellowstone wolf. 105, her sister, had died in June 2003. The third sister, 106, the alpha female of the Geode pack, was still going strong.

302 continued to show up in Lamar Valley. There was no sign of the gray female who had been traveling with him. 253 seemed to be on guard for the interloper. One morning I saw 253 stalking 302, who moved off, but not fast enough. 253 ran all out at the rival male. 302 raced to the Lamar River, waded out about ten feet, then turned to face 253. As 253 glared at him, 302 backed away through the water. Coming forward, 253 got to within a few inches of 302 and they sniffed each other. 302 backed away, turned around, and ran off. 253 pursued him through the shallow section of the river, then both males had to swim when they got to deeper water. 302 reached the far side of the river, ran out of the water, and continued to flee.

Later, when the Druids were at the Chalcedony Creek rendezvous site, 302 came into the area. He saw 253 and ran off with the Druid male in pursuit. 21 got up and watched his son chase the intruder, then joined him in driving off 302. As he had done once before, 302 gave the two Druid males the slip by running into a bison herd.

302 would not give up on trying to approach the Druids. Since the mating season was over, his persistence suggested he understood that some of the young Druids were his offspring. Later in the day, I saw 21 chasing him once more. 302 ran to the Lamar River and swam across. 21 jumped into the water and swam after him. 302 climbed out of the river, ran north, then bedded down near the road and looked back at 21. Later 302 tried to feed on a bison carcass that the Druids controlled, and once again 21 chased off his nephew. Their long-running contentious relationship seemed like it would never end.

In mid-April, I saw three Druids at the black bear den near their den forest. The head of the mother bear was visible in the opening at the base of the tree. A gray yearling moved closer, and the sow swatted at it. Another yearling approached, and the bear took a swing at that wolf too. After the wolves left the site, two tiny bear cubs came out of the den, wrestled with each other, then tumbled downhill. Later the cubs climbed up the trunk of the dead tree above the den. They could climb better than they could walk. The bear family soon left the site and we lost track of them.

I saw 255 right after that near the northern base of Junction Butte, about a mile northeast of Tower Junction. She had been in the area there for several days and we thought she was denning somewhere close by, as she no longer looked pregnant. Doug did a flight the next day and saw a rocky outcrop that seemed to be her den site. We had seen 255 mate with two Mollie's males and with a Leopold wolf, but none of them stayed with her. She would be a single mother.

At the end of 2002, there had been five packs in the former Druid superterritory. Now two more groups had formed there: U Black's Specimen Ridge pack and, assuming she had pups and they survived, 255's as-yet-unnamed single-parent family. They were trying to squeeze into an area that already had a high density of wolves. I expected that some packs, probably the smaller groups, would fail.

In late April, we often saw Druid wolves go up and over the Middle Foothill at the Chalcedony rendezvous site and disappear into the trees to the south, where U Black had denned the previous year. 376 had looked pregnant and it seemed like she was denning at that site. She had mated three times with a large black wolf that seemed to be from the Mollie's pack, but I had also seen her flirting with 302, so there was a chance he might have bred her as well.

One spring evening, as I was driving back home in the dark, I saw my headlights reflecting off the eyes of a tall animal on the road. I first thought it was a moose, but when I got closer, I saw that it was a grizzly standing up in my lane. I stopped and waited for it to move off the road.

21

May

I CHECKED ON THE Geodes' den site at the base of the cliff in the Hellroaring area in early May and saw 106, a gray adult, and six pups. Eventually eleven pups were counted there. Eleven is the highest known pup count for a mother wolf in Yellowstone, so some might have belonged to another female in the pack.

The next day, I saw 21 carry an elk leg back to the main Druid den forest, an indication that alpha female 286 had pups there. We later got a count of five pups at that site, but six were seen from the air. Former Druid and now single mother 255's signal came from that site early that morning. She then traveled back to her own den, fourteen miles to the west, arriving eight hours later. She was probably paying a social visit to her birth family.

A few days later, there was evidence that 376 had pups in the trees behind the Chalcedony Creek rendezvous site. A black yearling came into the area and 376 ran to it, licked its face repeatedly, and got a regurgitation. I also saw 21 going up into the trees. We spotted 302 regularly at the rendezvous site and he also went up to the den site. That made me wonder if he had bred 376 and thought that the pups there were his.

376 had a special radio collar that beamed her location to a satellite. Dan Stahler analyzed the data and found that from April 18 to April 26 the satellite recorded only one brief transmission, leading him to conclude that she was spending most of her time underground. Later 286 made regular visits to 376's den site at Chalcedony and the two mothers seemed to get along well.

As the Wolf Project continued to monitor 255, it appeared that her pups had not survived. She wasn't attending the site regularly as a mother with newborn pups would be doing. After she abandoned the site, we hiked out to the area and found her den under a boulder. Some of her fur was caught on the edge of that rock. We didn't find any dead pups in the den. Since 255 had no male wolf that regularly helped her, she probably hadn't gotten enough food to produce sufficient milk for her pups. In contrast, U Black, now the alpha female of the newly formed Specimen Ridge pack, was being supported by alpha male 194 and his gray brother.

The Buffalo Fork pack, the group started by 105, no longer had any collared wolves and we weren't seeing them from the ground. The surviving members had likely moved out north of the park onto National Forest land, so they were no longer considered a Yellowstone pack. That brought the number of packs in the former Druid superterritory back to five, the same as at the end of 2002.

The Slough wolves were denning up Slough Creek, near McBride Lake. Former Mollie's wolf 261 was still the alpha male, and the uncollared black who was probably a Druid was the alpha female. There were twelve adults in the pack, including three Mollie's.

I saw the first Specimen Ridge pup at U Black's den on May 11 and we eventually counted five pups there. One morning alpha male 194 bedded down and the pups climbed up on him. Two of them played together on their father's back. We often saw U Black nursing her pups there.

The Specimen Ridge adults were doing a good job of caring for and protecting their pups. I saw a black bear approaching the den area and the gray male chased it off. The bear climbed a tree close to the den and lay on a big branch. U Black bedded down fifteen feet from the tree, ready to drive off the bear when it came down. One of her pups ran over and the mother wolf got up right away, went to the tree trunk, and stood up against it on her hind legs, making sure the bear stayed up in the tree while that pup was nearby. After the pup moved off, the bear tried to come down the tree three times and the mother drove it back up on each occasion. When later it came down to the ground, U Black chased it away.

302 often returned to Lamar Valley and seemed especially interested in the den at the Chalcedony rendezvous site. On May 30, wolf watcher Marlene Foard saw 376 with two black pups. I joined Marlene on the hillside north of the road, across from that area, and she pointed out the mother and two very small pups. We saw them walk under the belly of a yearling wolf and there was three inches between the heads

of the pups and the yearling's belly. The next morning, I saw a pup nursing on 376. Later a male yearling regurgitated twice to the mother and pups. When 376 went to a nearby carcass, the yearling stayed behind and babysat the pups. That evening I saw 376 carry a pup around in her mouth after it had strayed too far away.

Shannon Barber continued her study on elk calf survival and got similar results to those of the previous year: bears were killing more calves than wolves. She had radio-collared eleven newborn calves. In late May, only three were still alive. Four had been killed by bears, one by wolves, one by coyotes, and one by either a bear or a wolf. She couldn't determine a cause of death for the eighth dead calf.

22

June

IN EARLY JUNE, I saw 21 at the Chalcedony Creek rendezvous site. He was with the Druid alpha female, 286, who must have been taking a break from her five pups back at the main den. 376 was also there with her two black pups. 286 went to the pups and gently interacted with them. Some bison came into the area and the pups immediately disappeared into the complex of burrows at the site.

Now that 255 had lost her pups, she was back with the Druids and helping the two mother wolves. 376's pups tried to nurse on 255. She must have been still producing milk, for she stood in place and didn't seem to mind having the pups there. After that 255 bedded down. The pups climbed up on her back and wrestled with each other. We saw that one pup was male and the other female.

The next morning, I saw a black male that I didn't recognize having a friendly interaction with the Druids at the rendezvous site. He had a white blaze on the lower part of his chest and the skinny appearance of a yearling. The Druid yearlings jumped around him playfully and 376 licked his face, then pawed at his head. He wagged his tail at her. The new male walked over to one of the pups. Mother 376 seemed to have no problem with that. He lowered his head and sniffed the pup. The unknown black hung around the site with the Druids for the next two days. 21 and 253 didn't visit the site during that time.

I got up at 3:20 a.m. on June 11 and left my cabin in the dark at 4:37 a.m. New snow was on the ground and the temperature was 37 Fahrenheit (3 Celsius) at our high-elevation mountain town. As I pulled out of my driveway, I didn't know this day was going to be a major milestone in the story of the Yellowstone wolves.

I walked up through the snow on the hillside north of the road and at 5:47 a.m. looked out at the Chalcedony rendezvous site. I spotted a black wolf bedded down there. Then 376 came in from the west. Her two pups ran out to her and she greeted them. 253 and one of the gray yearlings were nearby, traveling to the west. I spotted a fresh bull elk carcass past 253 on the south bank of the Lamar River. 21, 286, and two yearling Druids were at or near the carcass.

Soon 21 joined 376 at the rendezvous site. She later went to the elk carcass and fed. 286 crossed the road and returned to her pups at the main den. Things slowed down after that. In late morning, I headed in for a break and to work on my field notes, then started back out at 6:56 p.m. At that time of the year, we could watch wolves until nearly 10:00 p.m.

I got in place on the hill and saw Druid wolves bedded down at the Chalcedony rendezvous site. 376 went to 21 and licked his face. Her two pups and other wolves came over and all wagged their tails at 21. His daughter 255 rolled on her side next to him.

255 and 286 started a howl and 21 joined in, as did the other adults and pups. More Druids ran in and greeted the wolves in the main group. The pups then wrestled each other next to 21. He didn't play with them, and I thought it was just because he was tired. I then realized that I hadn't seen him play much with the yearlings and young pups in recent months. In past years, he seemed to go out of his way to interact with the pups.

I saw a bull elk to the northeast of the wolves at 8:36 p.m. 255 got up and stalked in his direction. The bull ran east, and she took off after him. Two other bull elk joined him. A black male yearling joined 255 in chasing the three elk.

I swung my scope back to 21, expecting to see him racing toward the other wolves. Instead he was sitting up and just watching the chase. That was not like him. He should have been helping the two younger wolves pull down one of the bulls. It was like he had no energy left in his old body. He was over nine years old and likely the oldest wolf in the park, the last of his generation.

As I looked at 21, I realized how thin and gray he was getting. Since I saw him most days, I didn't notice the gradual change in his appearance, but when regular visitors returned to the park after a significant absence, they commented on how he had aged. The graying process seemed to accelerate after the loss of 42. When I recently looked at photos of him taken during that time, I was shocked at how old he looked.

I turned my scope to the pursuing wolves and saw 255 nipping at the hind legs of the biggest bull. He stopped, turned around, and faced her. The bull was probably about 700 pounds while 255 was well under 100 pounds. She was overmatched and needed help. One of the gray yearlings ran in, but she was also a small wolf and the two females would have little chance against such a big bull. 255 went after a smaller bull, but he outran her. She stopped and looked back to the west, toward 21. He was still just watching.

The gray yearling was now chasing the big bull and gaining on him. She caught up and ran beside him, looking for an attack point. 255 sped toward them. The elk was now heading toward the Lamar River and seemed to be tiring out. As he slowed down, both females were right with him. 255 nipped at his hind legs and he kicked back at her. 255 bit into a hind leg but lost her grip. The two females needed 21 to come in and use his great strength to grab the bull by the throat and wrestle him down.

The elk reached the riverbank, jumped down, and splashed into the water. He swam across to the north, climbed out on the far side, and ran west. Determined to get him, the gray yearling paralleled him along the south bank. Farther west, the bull recrossed the river to the south and continued running west. He was exhausted now, and the wolf caught up with him. She bit at a hind leg but had to let go when he kicked back at her. Then she bit higher up on that leg. After that she got two bites into his rear end.

The bull kicked back again and hit her full on. She got knocked to the ground but jumped right back up and courageously resumed the chase. Catching up, she grabbed a hind leg but had to let go to dodge a kick. The bull made it back to the river and ran into the water. He then stopped and turned to face the wolf. The gray yearling waded out through chest-deep water and stood in front of him. He towered over her. She was outmatched and desperately needed other wolves to help her out. I looked around and did not see 21. I couldn't understand why he hadn't charged in to help his daughter. It was now nearly 10:00 p.m. and I had to head in because it was getting dark.

June 12 was the fourth anniversary of my getting up early every day and going to study wolves. That added up to 1,461 days in a row. I saw a group of Druids at the rendezvous site, but not 21. His radio-collar battery no longer worked, so I had no way of locating him.

The black male with the white blaze came in from the east and once again had friendly interactions with the young adults. Soon after that, I saw 302 to the west. He fed on the elk carcass in that area, then continued east. The black male casually walked up to 302 with his tail in a neutral position. The two came together and seemed to know each other, suggesting that the black, like 302, had dispersed from the Leopold pack. When the new black trotted away, 302 followed. The two Druid male yearlings joined them. All three treated 302 like a higher-ranking wolf. 255 came over and they all had a friendly meeting.

The new black, 302, and the rest of the group arrived at the rendezvous site and met up with 376 and other young Druids. The two pups came into sight. When they saw 302 and the new black male, they disappeared into a nearby burrow. 302 walked over and looked into the entrance. The pups soon came out and tried to nurse on their mother. They still seemed wary about the two outsider males. 302 approached one of the pups. After sniffing it, he gave it a light holding bite on the back. The mother wolf immediately charged at 302 and he backed off. Both pups ran down into the burrow and 302 looked down the opening again. All the wolves then bedded down. I went back to the main den area and got signals from 253. I hadn't seen 21 so far and hoped that he was up there with 253.

When I got back out later in the day, 302 and the new black male were still at the rendezvous site. 253's signal continued to come from the main den. That meant he didn't know the two males were at the rendezvous site. The pups were still concerned about the two males and slipped into the burrow when they saw them approach. Later in the evening, the new black led east and most of the other adults followed. I lost them going up the trail to Specimen Ridge. We hadn't seen 21 that day and didn't know where he was.

That black male would later be collared and given the number 480. He turned out to be a nephew to 302. He was born in 2003, three years after his uncle, and later proved to be a much tougher wolf than 302. Soon 302 and his nephew would gang up on 253 and try to take over the Druid pack, but that is a story for another book. This story is about 21.

23

The Quest

As the days went by, 21 continued to be missing. A month soon passed with no sightings of him. His long absence suggested something was very wrong.

Wildlife research is often like being a detective: you try to figure out what happened and why. I went back through my field notes, all the way to late January, to look for clues to help me understand why 21 left his family.

42 was last seen late in the day on January 31. The next morning, 21 was alone several miles west of the other Druids, who were at Slough Creek. He went to a fresh bison calf carcass and fed on it. Later the younger Druids joined him there. The wolves at the carcass howled frequently, perhaps trying to contact the missing 42.

We saw the Mollie's wolves to the south, on top of Specimen Ridge. One of them had blood on his muzzle and a fresh wound in his side. That was our first clue that the two packs had fought during the night, probably at the site of the bison carcass.

As I mentioned earlier, I imagined 21 fighting with several rival males while other Mollie's wolves went after 42. She likely ran south across the road with wolves pursuing her. That direction would take her to the base of Specimen Ridge. From there she went to the top of the ridge, then on to where Dan Stahler saw her body and signs of a battle.

On the morning of February 2, we saw 21 south of the road in the Crystal Creek area. He was alone and howling, seemingly trying to locate 42. On the third, 21 and the Druids went back to their main den in Lamar Valley. That would be a likely place to look for 42.

I studied my field notes from the rest of February and March to check on where the Druids had been seen. I found that during that period they had ranged throughout their entire territory, from Round Prairie, five miles east of their den site, to Tower Junction, fifteen miles west of the den. I saw them on the north and south sides of the road throughout that section of the park. As I thought about all that traveling, I wondered if 21 was searching those areas for 42.

There were three occasions during those weeks when the Druids came close to finding 42's remains. In early March, I saw the pack bedded down on the crest of Specimen Ridge, above the Crystal Creek drainage. 21 got up and traveled farther west, but then stopped and bedded down once more. He unknowingly was about three miles from where 42 had died. The next day, the pack traveled west from there and got to within two miles of the site of her death, then got distracted by a band of bighorn sheep. The Druids were up on Specimen Ridge again on March 26, this time to the south of Slough Creek. I saw 21 lead them over the crest, but they soon turned around and came back down to the valley.

286 was due to have her pups around April 7. She would have led the pack back to the Druid den forest in Lamar, and 21 and the other wolves would have to be based there during the coming months. I visualized 21 waiting for 42 to come back to that den site, the most likely place for her to reappear. She never arrived, and all the Druids had to devote themselves full-time to supporting the pack's two mothers and their pups during the next two months.

By early June, it had been four months since the loss of 42, but I wondered if, like elderly humans, 21's sense of time might have been distorted and perhaps compressed. Did he feel it was a much shorter time span?

I last saw 21 on the evening of June 11. In mid-July, an outfitter found the remains of a collared wolf at the Druids' Opal Creek rendezvous site. He took the wolf's collar and gave it to backcountry ranger Mike Ross, who notified me. I met up with Mike and he handed me the collar. On the inside of the leather strap there was a faded radio frequency. It was the one assigned to 21. I called Doug at the office and filled him in on what we had learned.

On the morning of July 23, Doug Smith, Dan Stahler, Matt Metz, Emily Almberg (my co-worker that summer), Monty Simenson, a Park Service cowboy, and I met in the Hitching Post parking lot, across from the Druid den forest. We got on our horses, took the Specimen Ridge trail uphill, then veered off to the east and arrived at the Opal Creek rendezvous site just after 1:00 p.m. The outfitter had told us where to look for 21.

We dismounted and walked up the low hill on the east side of the meadow and saw his remains. It appeared he had bedded down there and died peacefully in his sleep.

After I last saw him, I could imagine him using the last of what strength he still retained and going up the same route we had taken to the meadow, then lying down on that hill, in the shade of a nearby lone tree.

All of us had known 21 well and none of us could find a way to verbalize our feelings. Doug and Dan took 21's skull for the park's museum collection. Like 8 when he was old, 21 had suffered damage to his jaw. One of his four canine teeth was broken off and two other canines were chipped. Only one was intact. His other teeth were in good shape for an old wolf.

After that we walked separately around the area. I saw fresh elk beds in the lush meadow grass. There was a tributary of Opal Creek running through the area, and a small pool of water was in another section of the meadow. The spring wildflowers were still in bloom this high up. The dominant species was forget-me-not.

This was the site where Bill Wengeler and I had found the Druids when we hiked up on Specimen Ridge in September 1999. From a distance, we had seen 21, 42, and the rest of their family bedded down in that meadow, and I remembered seeing that shade tree on the hill.

Doug spoke of frequently seeing the Druid adults and pups in this meadow over the years on his tracking flights. It was a favorite late-summer, high-elevation rendezvous site for the family. 21 and 42 likely bedded down together on that hill many times and watched their pups and yearlings play in the meadow. They would have been like a human couple who had a cabin in the mountains, watching their kids playing from their porch.

To me this was a sacred site for 21 and the Druids and I felt a like an intruder, so I was glad when Doug suggested we ride off. We got back to the Hitching Post lot three hours later.

As I drove out of the lot, I saw one of the Druid yearlings coming from the den forest and crossing the road to the south. We called that spot 21's Crossing because he so often left the den on hunts through that area and returned on the same route to bring food to the pups.

On the way home, I thought of the bright blue forget-me-nots that surrounded the remains of 21. That was the moment when the deep emotions of his loss really hit me. I was saddened that his life was over, but grateful that he was able to reach that meadow and got to lie down to rest on the hill. As his remains further decomposed, they would provide a bit of nourishment for future generations of forget-me-nots, a fitting tribute to his life and legacy.

I spent a lot of time thinking about 21 over the years and why he went up to that meadow. For a while, I thought that it might have been like a dying king in Scotland, back in the time of Druid priests, climbing a peak in the Highlands to take one last look at the territory he ruled. But then I realized that 21 would not have done it for that reason. My take on him was that he was never impressed by the size of his pack, the extent of his territory, the rivals he had defeated, or the battles he had won. 21 was a warrior king who never had a crown, and that insignificant hill was the closest thing he had to a throne room.

If 21 didn't climb that mountain to see the extent of his territory, why did he make such an arduous journey at the end of his life, when his great strength was rapidly diminishing? I think it was because 21 still didn't know 42 was dead, only that she was missing. He had not found her during his extensive travels throughout the Druid territory so perhaps he decided to use his final days to go on a quest, to visit one last spot, the Opal Creek rendezvous site, and look for her there. That is the meaning of such a site: it is a place to meet up.

She wasn't in that meadow, but 21 would have sniffed the lone tree and gotten her scent on the trunk from the many times she and 21 had marked it. It would have been old, but it was her scent. At least he had that.

Can a wolf feel happiness and joy? I think 21 did at that moment.

I imagined him then walking the few feet to the hilltop, where he had bedded down so many times next to his life partner, and lying down to rest. As he slowly drifted off into sleep, I would like to think that the scent from that tree triggered a picture. If so, then the last thing in 21's mind as he lost consciousness for the final time was an image of 42.

EPILOGUE

IN THE SMALL town of Gardiner, just outside the North Entrance to Yellowstone, the nonprofit organization Yellowstone Forever has a visitor center and store. Inside the building there is an exhibit room with displays about the park. A statue is among the exhibits.

That statue could have depicted any one of the many prominent individuals who played major roles in the history of Yellowstone. It could have been President Teddy Roosevelt, one of the early National Park Service directors such as Stephen Mather, a renowned park superintendent like Horace Albright or Mike Finley, an accomplished wildlife biologist like Doug Smith, or a ranger who had served the park in extraordinary ways. It might have been a famous park grizzly such as Scarface or a legendary alpha male like 8 or 21.

That statue is of none of those worthy candidates. It is a statue of wolf 42.

ACKNOWLEDGMENTS

I FIRST WANT TO thank my editor, Jane Billinghurst, for working way beyond the call of duty to make my original manuscript far more readable and polished than it would have been without her help. Thanks also to Rob Sanders at Greystone Books for accepting my proposal for this series. Thanks to Rowena Rae for doing the copyediting and to Meg Yamamoto for proofreading the final version of the book. I also want to thank Fiona Siu and Nayeli Jimenez for designing the look of the book and all the hardworking people who helped get it out onto bookstore shelves and into the hands of readers. Everyone has been very supportive and encouraging. Thanks also to Yellowstone National Park for all it has done for me, for the wolves, and for the millions of people who visit the park every year.

My good friend Laurie Lyman read the first draft of the book and gave me very helpful comments and suggestions on how I could improve it.

There were many current and former National Park Service staff members and wildlife researchers who advised me on their work and experiences with wolves. Here are the ones who were especially helpful: Norm Bishop, Jim Halfpenny,

Bob Landis, Rolf Peterson, Dan Stahler, and Jeremy Sunder-Raj. I want to give special thanks to Kira Cassidy, who drew and illustrated the maps for the book. Scores of volunteers have worked for the Wolf Project over the years and every one of them was very helpful to me. Thanks also to the photographers who allowed me to use their pictures in the book.

Special appreciation goes to Doug Smith, the lead biologist for Yellowstone National Park's Wolf Project. Doug is a unique PhD-level scientist who can relate what he has been learning about wolves and the natural world to regular people in ways that not only educate but, more importantly, inspire them. Despite Doug's heavy workload and obligations to his family, he took the time to read my manuscript and suggested valuable changes and additions. They made this book much better than it would have been without Doug's generous involvement.

I want to make special mention of two couples who made a big difference to my life: Gary Hoshiyama and Patricia Wood, and Bob and Annie Graham. There have been hundreds of wolf watchers who have greatly aided me over the years. On many occasions when I was trying to find wolves, another person would spot them first and graciously point them out to me. I would also like to thank the vast numbers of Yellowstone visitors who have been kind and friendly to me over my many years here. There is something about being in the park that seems to make people positive and sharing. Thank you to everyone I have met over the years. I could not have done this without all of you. I regard this book as a joint effort.

AUTHOR'S NOTE

INSPIRED BY HOW members of a wolf pack support each other in time of need and by all the kind people who have helped me over the years, I will be donating proceeds from my wolf books to Yellowstone National Park and to non-profit organizations such as the Make-A-Wish Foundation of America and the American Red Cross. Readers who are interested in helping support wolf research and wolf education in Yellowstone can go to the Yellowstone Forever website, www.yellowstone.org/wolf-project, to make a donation. Yellowstone Forever is the official nonprofit partner of Yellowstone National Park and helps fund the park's Wolf Project, the National Park Service operation I used to work for.

REFERENCES

Grandin, Temple. 1995. *Thinking in Pictures and Other Reports from My Life with Autism*. New York: Vintage Books.

Haber, Gordon, and Marybeth Holleman. 2013. *Among Wolves: Gordon Haber's Insights into Alaska's Most Misunderstood Animal*. Fairbanks, AK: University of Alaska Press.

Halfpenny, James. 2012. *Charting the Yellowstone Wolves: A Record of Restoration*. Gardiner, MT: A Naturalist's World.

Heinrich, Bernd. 1989. *Ravens in Winter*. New York: Summit Books.

Israelsen, Brent. 2001. "Beloved Wolf, 253, Running with Original Pack." *Salt Lake Tribune*, December 21, 2001.

Israelsen, Brent. 2002. "Wandering Wolf Is Well Known to Yellowstone Visitors." *Salt Lake Tribune*, April 12, 2002.

Kennedy, Des. 1988. "The Great Transformer." *Nature Canada*, Summer 1988.

Kipling, Rudyard. 1895. *The Second Jungle Book*. London: Macmillan and Co.

Mech, L. David, Douglas W. Smith, and Daniel R. MacNulty. 2015. *Wolves on the Hunt: The Behavior of Wolves Hunting Wild Prey*. Chicago: University of Chicago Press.

PBS. 2014. *Inside Animal Minds*. NOVA series, April 23, 2014.

Smith, Douglas, and Gary Ferguson. 2005. *Decade of the Wolf: Returning the Wild to Yellowstone*. Guildford, CT: Lyons Press.

Smith, Douglas W., Kerry M. Murphy, and Debra S. Guernsey. 2001. *Yellowstone Wolf Project: Annual Report, 2000*. Yellowstone National Park, Wyoming: National Park Service. YCR-NR-2001-02.

Smith, Douglas W., and Debra S. Guernsey. 2002. *Yellowstone Wolf Project: Annual Report, 2001*. Yellowstone National Park, Wyoming: National Park Service. YCR-NR-2002-04.

Smith, Douglas W., Daniel R. Stahler, and Debra S. Guernsey. 2003. *Yellowstone Wolf Project: Annual Report, 2002*. Yellowstone National Park, Wyoming: National Park Service. YCR-NR-2003-04.

Smith, Douglas W., Daniel R. Stahler, and Debra S. Guernsey. 2004. *Yellowstone Wolf Project: Annual Report, 2003*. Yellowstone National Park, Wyoming: National Park Service. YCR-NR-2004-04.

Smith, Douglas W., Daniel R. Stahler, and Debra S. Guernsey. 2005. *Yellowstone Wolf Project: Annual Report, 2004*. Yellowstone National Park, Wyoming: National Park Service. YCR-2005-02.

Stahler, Daniel R. 2000. "Interspecific interactions between the common raven and the gray wolf in Yellowstone National Park, Wyoming: Investigations of a predator and scavenger relationship." Master of science thesis, University of Vermont.

Stahler, Daniel R., Bernd Heinrich, and Douglas Smith. 2002. "Common ravens, *Corvus corax*, preferentially associate with grey wolves, *Canis lupus*, as a foraging strategy in winter." *Animal Behavior* 64(2): 283–290.

Wilmers, Christopher C., et al. 2003. "Resource dispersion and consumer dominance: Scavenging at wolf- and hunter-killed carcasses in Greater Yellowstone, USA." *Ecology Letters* 6: 996–1003.

Yellowstone National Park. 2017. *Yellowstone Resources and Issues Handbook: 2017*. Yellowstone National Park, Wyoming: National Park Service. https://www.nps.gov/yell/learn/resources-and-issues.htm.

INDEX

Maps and wolf charts indicated by page numbers in italics

affection, expressions of, 62

Agate Creek wolf pack: confrontation with wolf 302 and companions, 190–91; denning sites and offspring, 114–15, 165; encounters with Druid Peak pack, 100–101, 117–20, 122; formation of and growth, 96–98, 99–100, 103–4; membership, *94*, 140, *152*; pup count, 128, 171; territory, *92*, *150*, *196*

—wolf 103: Agate Creek pack formation and, 99, 103–4; assistance for wolf 42 with pups, 15–16, 17, 18, 19; assistance from wolf 42 with pups, 75; consolidation of Druid pups at main den and, 28; cooperative behavior, 46; death, 190, 216; departure from Druid Peak pack, 80; elk hunting, 20; female hierarchy within Druids and, 64–65; first den and litter, 67, 68–69, 70–71, 74, 75, 76, 78; mating, 58, 59, 60, 99, 114–15, 165; new den site, 71; New Gray and, 40; play, 46; radio collar, 57; Tower pack encounter, 20–21

—wolf 113 (alpha male): with alpha female, plate 6; attitude toward rivals, 185; background and leadership style, 98; encounters with Druid Peak pack, 101, 119–20, 122; mating and offspring, 99, 114, 117, 118, 165; pack formation, 96–98

—wolf 251: death, 173, 187; denning with Agate pack, 170; dividing time between Agates and Druids, 115–16, 117–18;

mating and pregnancy, 100–101, 165; parentage, 171; pups, 171, 187, 191
—wolf 472 (alpha female), 115, 117–18, 165, 170, plate 6
Almberg, Emily, 232
alpha pair, 29–30, 54–55, 62
Amethyst Creek, 182
Andre, Melissa, 13
antlers, 188
aspen, 126

Babbitt, Bruce, 56
babysitting, 19, 74–75, 122, 173
Bangs, Ed, 145
Barber, Shannon, 172, 223
Barron, Eric, 34–35
bears. *See* black bears; grizzly bears
Beartooth wolf pack, 44
Beattie, Mollie, xv
beavers, 71–73, 126
Bechler wolf pack, territory, *92*, *150*, *196*
beds, in snow, 7
birds: Canada geese, 79; northern harrier, 79; ravens, 14, 130–35; sandhill crane, 181; songbirds, 127
Biscuit Basin wolf pack, *196*
bison, 13, 39, 109–10, 163–64, 179, 180
black bears, 116–17, 215, 218, 222
Blacktail Plateau, 47
breeding, 10, 54–55, 60, 101, 114
Brown, Cliff, 82–83
Buffalo Fork wolf pack: confrontation with Rose Creek pack, 173; formation of, 99, 103–4, 140; loss of pups, 128; mating, 101; membership, *94*, 140; move outside Yellowstone, 221; sightings of, 127; territory, *92*, *150*
—wolf 105 (alpha female): aggression from wolf 40, 21; Buffalo Fork pack formation and, 103–4; consolidation of Druid pups at main den and, 27, 28; cooperative behavior, 29; death, 172–73, 216; den appropriated from Rose Creek pack, 66; den shared with wolf 42, 15–16, 17, 18, 19, 25, 29; departure from Druid Peak pack, 80; elk hunting, 20; female hierarchy within Druids and, 65; mating and pups, 10, 29, 58, 67–68; New Gray and, 40, 58; pup rearing, 39, 40; radio collar, 57
—wolf 218, 99
bullying, 24, 90

Cache Creek, 41
Canada geese, 79
Carter, June, 204
Cash, Johnny, 204
Chalcedony Creek rendezvous site, 21, 38
Chan, Brian, 31, 100, 102

Chief Joseph wolf pack: membership, 98; territory, *2, 52, 92, 150, 196*
—wolf 113. *See under* Agate Creek wolf pack
Chin, Susan, 77
cooperative behavior, 29, 45–46, 49–50, 75–76
Cougar Creek wolf pack, territory, *52, 92, 150, 196*
cougars. *See* mountain lions
coyotes, 41, 114, 121, 170–71, plate 3
Crystal Creek, 126
Crystal Creek wolf pack. *See* Mollie's wolf pack

Dance, Doug, 192
Dead Puppy Hill, 32–33
Death Gulch, 41–42
deaths, 141–42
Denali National Park, 34, 146–47
dens, 68, 71
disabilities, 85–86
Druid Peak wolf pack: attack by Geode Creek pack, 105–8, 109, 111; attack by Nez Perce pack, 81–85; attacks by Mollie's wolves, 136–38, 139, 205–9, 230–31; babysitting by yearlings, 74–75; background, xv–xvi; beaver encounter, 71–73; bird encounters, 79, 181; bison hunting, 179, 180; black bear encounters, 215, 218; coexistence with Slough Creek pack over bison carcass, 183–85; cooperative behavior, 29, 45–46, 75–76; dead pup, 182; den locations, 63, 65–67; dispersal to form new packs, 60, 80, 99, 103–4, 140; elk hunting, 14, 17, 18–19, 20, 31–32, 37, 48, 55–56, 57, 71, 89–90, 103, 109–10, 130–31, 168–69, 187, 226–28, plate 2; encounters with Agate Creek pack, 100–101, 117–20, 122; encounter with Rose Creek pack, 89; expansion into Rose Creek pack territory, 45, 47, 55, 61, 88–89; grizzly bear encounters, 35, 36–37, 39, 41, 44, 75, 76–77, 116–17, 167–68; hierarchy changes, 46, 64–65; insurrection against wolf 40, 24; leadership, 29–30, 110–11, 209; membership, *3–4*, 6, *53*, *93*, 128, *151*, 154, *197*, 212; moving pups to Chalcedony Creek, 38; mule deer hunting, 13; outside suitors for females, 96–97, 201–2, 203; play, 12, 14, 36, 46–47, 73, 79, 85, 102–3, 121–22, 125, 161, 181; potential competition from Nez Perce pack, 61; pup consolidation, 26–28, 30, 177–79; pup counts, 33, 35–36, 67–68, 113–14, 173–74, 177–78, 200,

220; pup rearing, 38–42, 43, 116–17; pup survival rate, 36, 49, 79, 123, 154, 181; radio collars, 9–10, 57, 154–55, 200; reunion at Chalcedony Creek, 76; road crossings, 17–18, 32–33, 45; road safety lessons for pups, 34–35, 37, 100, 102; scavenging from mountain lion, 189; size record, 74, 78–79; swimming lessons for pups, 120–21, 175–76, 178; territory, *2*, *52*, 55, *92*, *150*, *196*; wolf 302 courtship of Druid females, 155–59, 160, 162–63, 185–86, 192, 200–201, 202, 216, plate 7

—Half Black: consolidation of pups and, 177–78; den site, 165; disappearance of, 212; encounter with Slough Creek pack and, 183; in female hierarchy, 161; mating, 203; naming of, 131; pups, 166; wolf 302 and, 156, 159, 160, 162–63, 177, 186, 192, 193, 202

—New Gray: aggression from wolf 21 toward, 54, 60, 80; care for pups, 40–41, 76; departure from Druid Peak pack, 80; in male hierarchy, 46; mating attempts, 57, 58, 60; maturation, 42; parentage, 111; play, 46–47; as possible Geode Creek alpha male, 111; Saddleback and, 42, 59. *See also* Geode Creek wolf pack—wolf 294 (alpha male)

—Saddleback: assistance for wolf 42 while denning, 17, 18; elk hunting, 14, 48, 57; grizzly bear encounter, 35; leadership experience, 43; in male hierarchy, 46; maturation and departure from pack, 57, 58–59, 59–60; mule deer hunting, 13; naming of, 6; New Gray and, 42, 59; play, 12, 14; vole hunting, 7; as yearling, 16

—Stripe: assistance for wolf 42 while denning, 17, 18; death, 31; elk hunting, 14; mule deer hunting, 13; naming of, 6; play, 12, 14; as yearling, 16

—U Black. *See under* Specimen Ridge wolf pack

—wolf 21 (alpha male): advancing age, 125, 226–27, 228, plate 1; affection with wolf 42, 10, 45, 46, 62, 125, 189, 201, 203, plate 8; attack by Geode Creek pack and, 105–7, 108, 109; attack by Nez Perce pack and, 81–82, 83; attacks by Mollie's wolves and, 137–38, 207, 208; attempt to re-create last months of, 230–32; attitude toward rivals, 185; babysitting by, 173; background, xv–xvi; break from pups, 45; care for wolf 40's pups after her death, 24–25, 25–26; character, 65, 83, 108; comparison to wolf 302, 160–61, 191; confrontations with outside male suitors, 96–97, 156–58, 177, 186, 192, 201, 202, 211–12, 215–16, 217; coyotes and, plate 3; death, xiii, 230, 232–35; elk hunting, 14, 17, 32, 71, 103, 168; empathy for younger wolves, 90, 129; encounter with Agate Creek pack, 118–19, 119–20; encounter with Slough Creek pack, 183, 184; food provision, 17–18, 35, 37–38, 39, 40, 66, 171, 179, 220; grizzly bear encounters, 35, 36–37, 76–77, 168, plate 5; impact of, 42; lessons from, 138–39; licking wolf 253's paws and face, 214–15; male hierarchy and, 46; mating, 8, 10–12, 54, 56, 57–58, 59, 60, 99, 112–13, 159, 160, 202, 203, 209; New Gray and, 41, 42, 54, 60, 80, 111; play, 12, 14, 46–47, 73, 85, 102–3, 125, 161, 181; as pup, plate 1; pups raised by, 43, 166; pup training, 102–3; radio collar, 57, 166; relationship with new alpha female, 214; road crossings, 17–18, 32; running with pack, plate 4; scent recognition, 86; treatment of pups from other males, 214; wolf 42 as alpha female and, 29–30, 110–11; wolf 42's death and, 212

—wolf 38 (alpha male), xv, xvi

—wolf 40 (alpha female): aggression toward sister and other females, 11, 13–14, 16, 21, 29; background, xvi; den site, 15; elk hunting, 14; insurrection against and death, 22–24; interference in wolf 21's mating, 10, 11; mating, 8, 11–12; play, 12; radio collar, 9–10

—wolf 42 (alpha female): advancing age, 125, plate 4; affection with wolf 21, 10, 45, 46, 62, 125, 189, 201, 203, plate 8; assistance from others with pups, 15–16, 17, 18–19; assistance with other's pups, 66, 75; attack by Geode Creek pack and, 106–7, 108; attack by Mollie's wolves and, 137–38; attack by Nez Perce pack and, 81–82; background, xvi; beaver encounter, 72; bedding in snow, 7; break from pups, 45; chasing away younger females, 80; confrontation with wolf 302, 159; consolidation of pups at main den, 26–28; cooperative example, 49; death, 206–7, 208–9, 230–31; denning and pups, 17, 19–20, 63, 65–66, 112–14, 166; den shared with wolf 105, 29; den sites, 15, 68, 165; elk hunting, 18–19, 20, 32, 103; encounter with Slough Creek pack and, 183, 184; female hierarchy within Druids and, 30, 59, 64–65; food provision for pups, 122, 179; insurrection against sister, 24; leadership by, 110–11; mating, 8, 10, 56, 57, 99, 113, 159, 160, 202, 203; as new alpha female, 29–30; New Gray and, 41; play, 47, 73, 125; pup consolidation at rendezvous site, 38, 177–79; pups saved from sister, 25; radio collar, 57, 166; running with pack, plate 4; Saddleback and, 60; sister's aggression toward, 11, 13–14, 16, 21, 29; statue of at Yellowstone Forever, 236; swimming lessons for pups, 38, 120–21, 178

—wolf 103. *See under* Agate Creek wolf pack

—wolf 105. *See under* Buffalo Fork wolf pack

—wolf 106. *See under* Geode Creek wolf pack

—wolf 217. *See under* Slough Creek wolf pack

—wolf 218. *See under* Buffalo Fork wolf pack

—wolf 224, 90, 215

—wolf 251. *See under* Agate Creek wolf pack

—wolf 253: attempted takeover from Leopold males and, 229; babysitting duties, 173, 183; as beta male, 101; bone shared with pup, 181; confrontations with outside suitors, 97, 217; defense against Geode Creek pack, 107, 108, 111; elk hunting, 109–10, 168–69; grizzly bear encounter, 168; improvised running style, 109; injury from Nez Perce attack, 84–85; journey to Utah and back, 143–46, 172; play, 85, 122, 161; radio collar, 166; running with pack, plate 4; third leg injury, 188–89; walkabouts, 128, 210–11; wolf 21's licking of, 214–15

—wolf 254, 124

—wolf 255: attempts to help pups swim, 175–76; den site, 165; elk hunting, 226–27; encounter with Slough Creek pack, 183; in female hierarchy, 161; mating and pregnancies, 160, 166, 209, 210, 214; radio collar, 166; return to pack after loss

of pups, 224; single-parent family and loss of pups, 218, 221; visit to birth family, 220; wolf 302 and, 160, 163, 186, 192, 193, 202
—wolf 286 (alpha female): confrontation with Mollie's big black, 211; denning and pups, 220, 224, 232; encounter with Slough Creek pack, 183; mating, 207, 209; parentage, 154–55; radio collar, 166; sandhill crane chick and, 181; sightings of, 173, 177–78, 225; wolf 21's relationship with, 207, 209, 214, 224; wolf 302 and, 185–86; wolf 376 and, 221
—wolf 376: departure and return to pack, 210; mating and pregnancy, 211–12, 218–19; pups, 220–21, 222–23, 224; radio collar, 200; unknown black (wolf 480) and, 225; wolf 286 and, 221; wolf 302 and, 216

elk: antlers, 188; calving season, 31; den sites and, 68; hunting by Druid Peak pack, 14, 17, 18–19, 20, 31–32, 37, 48, 55–56, 57, 71, 89–90, 103, 109–10, 130–31, 168–69, 187, 226–28, plate 2; hunting by Leopold pack, plate 2; injuries from kick by, 63–64; lunges from bull elk, 187–88; migration to winter range, 46; misinformation about wolves and, 174; research project on calf predation, 172, 223; vegetation damage and restoration, 126–27; wolves' advantages when hunting, 48, 163–64
empathy, 90, 129

Finley, Mike, 56
Foard, Marlene, 222
food and feeding: nursing, 16, 17, 33; from people, 100; regurgitation, 17, 122; spoiled meat, 182–83. *See also* hunting
foxes, 116–17
Frazier, Scott, 129

Geode Creek wolf pack: attack on Druid Peak pack, 105–8, 109, 111; confrontations with wolf 302, 162–63; death of alpha male, 167; dens and pups, 170; formation of, 98, 103–4; membership, *95*, 140, *153*; pup count, 128, 220; radio collars, 160, 189; territory, *92*, 127, *150*, *196*
—wolf 106 (alpha female): attack on Druid Peak pack and, 107; care for wolf 40's pups, 25; consolidation of Druid pups at main den and, 28, 67–68;

continued good health, 216; cooperative behavior, 29; denning sites, 15, 66; departure from Druid Peak pack, 80; elk hunting, 31–32, 71, 89–90; female hierarchy within Druids and, 30, 64–65; Geode Creek pack formation and, 98, 103–4; mating, 11, 15, 58, 60; parentage, 30; pregnancy and pups, 165, 170; radio collar, 57, 160; Saddleback and, 59; vole hunting, 7–8
—wolf 294 (alpha male), 105–6, 111, 167. *See also* Druid Peak wolf pack—New Gray
—wolf 300 (alpha male), 167
Gibbon Meadows wolf pack, territory, *150*, *196*
Graf, Dan, 114, 116, 120
Graham, Annie and Bob, 155
Grandin, Temple, 87–88
grasshoppers, 39–40, 179
grizzly bears: cubs killed by males, 77–78; encounters with author, 33–34, 219; encounters with Druid Peak pack, 35, 36–37, 39, 41, 44, 70, 75, 76–77, 116–17, 167–68, plate 5; in spring, 13
ground squirrels, 7
Guernsey, Deb, 123, 206
Gunther, Kerry, 77

Haber, Gordon, 146–47
Half Black. *See under* Druid Peak wolf pack
Hamblin, Bill, 77
Hare, Brian, 49–50
Hayden Valley wolf pack, *196*
Heinrich, Bernd, 133–34
Hellroaring Creek, 55
Hudson, Tim, 190
hunting: for beaver, 73; for bison, 163–64, 179, 180; for elk by Druid Peak pack, 14, 17, 18–19, 20, 31–32, 37, 48, 55–56, 57, 71, 89–90, 103, 109–10, 130–31, 168–69, 187, 226–28, plate 2; for elk by Leopold pack, plate 2; for grasshoppers, 39–40; health issues in prey animals, 64; for voles, 7–8, 36, 39–40, 44–45, 76, 125; Wolf Project studies on, 12–13; wolves' advantages against elk, 48, 163–64

Indigenous peoples. *See* Native Americans
Inside Animal Minds (documentary), 49–50
Israelsen, Brent, 145–46

Japan, 129
Jimenez, Mike, 144, 172
Joshua Tree National Park, 204
joy, 47, 235

Junction Butte wolf pack, 54–55

Kendall, Wayne, 31, 171, 181
Kiowa Tribe, 127–28
Kipling, Rudyard: "The Law for the Wolves," 108–9

Lakota Elders, 67
Lamar River, 126
Landis, Bob, 128, 156, 158
Leopold wolf pack: aerial observation of, 89; affection between alpha pair, 47–48, 62; elk hunting, plate 2; fate of, 140–41; formation of, 89; pup count, 171; territory, *2, 52, 92, 150, 196*
—wolf 2 (alpha male), 47–48, 62, 89, 141
—wolf 7 (alpha female), 47–48, 62, 89, 140–41
—wolf 301, 159–60, 191
—wolf 302: attack on male rival, 213–14; attempt to take over Druid Peak pack, 229; comparison to wolf 21, 160–61, 191; confrontation with Agate Creek pack, 190–91; with gray female and two pups, 186–87, 188, 190, 192; offspring with Druid Peak females, 166, 200; other romantic relationships, 209–10, 212; parentage, 158; public perceptions of, 191–92; return visits to Druid Peak pack, 164–65, 165–66, 172, 176–77, 179, 180–81, 181–82, 192–93, 215–16, 217, 221, 222, 228–29; romantic relationships with Druid Peak females, 89, 155–59, 160, 162–63, 185–86, 192, 200–201, 202, 216, plate 7; subordination of, 160
—wolf 302's uncollared younger brother, 201, 202, 209, 210, 211
—wolf 480, 224–25, 228, 229
loyalty, 146–47, 215

MacNulty, Dan: *Wolves on the Hunt* (with Mech and Smith), 72–73
Make-A-Wish Foundation, 128–29
marriage partner, 204
marrow, 14
mating tie, 10. *See also* breeding
McIntyre, Rick: ancestry and surname, 202–3; assisting people to see wolves, 56–57, 67, 127–30; background, 5; cabin at Silver Gate, 171; early-morning observation streak, 33, 73, 173, 228; grizzly bear encounters, 33–34, 219; late-winter study, 12–13; lessons from wolf 21, 138–39; road crossing assistance for wolves, 32–33, 45;

technology and, 87; winter in Yellowstone, 5–6
Mech, Dave, 63, 64, 110, 172; *Wolves on the Hunt* (with Smith and MacNulty), 72–73
Metz, Matt, 141, 164, 170, 173, 207, 232
Mirror Plateau, 39
Mollie's wolf pack: background in Crystal Creek pack, xiv, xv, 136; bison hunting, 163; conflicts with Druid Peak pack, 136–38, 139, 205–9, 230–31; Slough Creek pack formed from, 139–40; territory, *2*, *52*, *92*, *150*, *196*
—wolf 194. *See under* Specimen Ridge wolf pack
—wolf 261. *See under* Slough Creek wolf pack
—wolf 377. *See under* Slough Creek wolf pack
—wolf 378. *See under* Slough Creek wolf pack
—wolf 379. *See under* Slough Creek wolf pack
mountain lions, 19, 44, 127, 169–70, 189
mule deer, 13
Murphy, Kerry, 9, 22, 39

Native Americans, 50, 67, 127–28, 129, 135
neophobia, 134
New Gray. *See under* Druid Peak wolf pack
Nez Perce wolf pack: attack on Druid Peak pack, 81–83; as potential competition for Druids, 61; territory, *2*, *52*, *92*, *150*, *196*
—wolf 214, 99–100, 114, 154–55
—wolf 215, 99–100, 101
—wolf 252, 100–101
northern harrier, 79
nursing, 16, 17, 33

Opal Creek rendezvous site, 40, 232–34, 235

partner, lifelong, 204
play: within Druid Peak pack, 12, 14, 36, 46–47, 73, 79, 85, 102–3, 121–22, 125, 161, 181; within Specimen Ridge pack, 222; with vole, 7–8
poisoned carcasses, 134
Potter, John, 129
pups: nursing, 16, 17, 33; regurgitated food for, 17, 122; resiliency and survival by themselves, 39–40, 55–56, 179. *See also* play

radio collars: for Druid Peak pack, 9–10, 57, 154–55, 200; explanation of, 8–9; fallen-off collars, 63, 73; for Geode Creek pack, 160, 189; life

saved by, 166; for Swan Lake pack, 155
Rathmell, Ray, 41, 157
ravens, 130–35; elk hunt and, 14, 130–31; poisoned carcasses and fear of potential dangers, 133–35; relationship with wolves, 131–33; Trickster stories, 135
regurgitation, 17, 122
Rickman, Carol, 11, 113, 171, 186, 205
Rickman, Mark, 11, 113, 171
road crossings: safety lessons for pups, 34–35, 37, 100, 102; by wolves, 17–18, 32–33, 45
Robertson, Don, 118, 119
Rose Creek wolf pack: confrontation with Buffalo Fork pack, 173; expansion by Druid Peak into territory of, 45, 47, 55, 61, 88–89; overview, 44; territory, *2*, *52*, *92*, *150*; Tower pack formed from, 20, 44
—wolf 8 (alpha male), xiv–xv, 63–64, 73–74, 158
—wolf 9 (alpha female), xiv–xv, 44, 66, 89
—wolf 18 (alpha female), 44–45, 111
Ross, Mike, 22, 23, 232

Saddleback. *See under* Druid Peak wolf pack
sandhill crane, 181
scent marking, 29–30, 42
scent recognition, 86
September 11, 2001, attacks, 79–80
Simenson, Monty, 232
Slough Creek, 73–74
Slough Creek wolf pack: aerial observation, 167; near Amethyst Creek, 182; coexistence with Druid Peak pack over bison carcass, 183–85; formation of, 139–40; howling, plate 6; leadership change, 192; membership, *95*, *152*, *198*, 222; retaliation against coyotes, 170–71; territory, *150*, *196*
—wolf 217 (alpha female): care for pups, 116, 122; confrontation with Druid Peak pack, 184; death, 201; encounter with Agate Creek pack, 117; loss of alpha status, 192; mountain lion encounter, 169–70; movement between packs, 99; pregnancy, 165; radio collar, 154; retaliation against coyotes, 170; Slough Creek pack formation and, 139–40
—wolf 261 (alpha male), 139–40, 162, 184, 185, 201, 222
—wolf 377, 203
—wolf 378, 203, 210
—wolf 379, 210
Smith, Doug: assisting people to see wolves, 56–57; bison hunt

by Mollie's wolves story, 163; den site investigation, 114; disabled wolf story, 86; on elk hunting misinformation, 174; investigation of wolf 21's death and, 232, 233, 234; investigations of remains of dead wolves, 124, 167, 207; McIntyre and, 87; passion for wolves, 86–87; radio collaring by, 9, 68, 154–55, 160, 189; road safety lessons for pups and, 100; September 11, 2001, attacks and, 80; tracking flights, 41, 47, 70, 75, 89, 121, 124, 170, 174, 190, 201, 218; on wolf 253's character, 146; *Wolves on the Hunt* (with Mech and MacNulty), 72–73

snow: bedding down in, 7; travel through, 49, 161–62

Soda Butte wolf pack. *See* Yellowstone Delta wolf pack

songbirds, 127

Specimen Ridge, 39, 40

Specimen Ridge wolf pack: formation of, 216, 218; membership, *198*; pup count, 222; territory, *196*

—U Black (alpha female): black bear encounters, 116–17, 222; denning and pups, 165, 166, 216, 222; in Druid Peak female hierarchy, 161; elk hunting, 130–31; food provision for pup, 179; mating, 99–100, 114, 115, 201–2, 209, 210, 212; Mollie's wolves and, 207; naming of, 131; support with pup rearing, 221; wolf 302 and, 156, 157–58, 159, 160–61, 162–63, 192, 193, 200–201

—wolf 194 (alpha male): pack formation by, 216; play with pups, 222; with U Black, 201–2, 209, 210, 212, 221

Stahler, Dan: investigations of remains of dead wolves, 167, 172–73, 232; research project on ravens and wolves, 133–34; tracking flights, 163, 171, 172, 179, 191, 206; on wolf 376's denning, 221; wolf genetic studies, 191

Stradley, Roger, 9, 45, 66, 78, 80, 211, 216

Stripe. *See under* Druid Peak wolf pack

Sucec, Rosemary, 67

Swan Lake wolf pack: background, 155; male suitor from, 202, 213–14; radio collars, 155; territory, *2*, *52*, *92*, *150*, *196*

swimming, 120–21, 175–76, 178

tail aversion, 10

thinking, in pictures, 87–88

Tower wolf pack, 20–21, 44, 92
Turner, Lisa, 190

U Black. *See under* Specimen Ridge wolf pack
Utah, 144, 145–46, 172

vegetation, restoration of, 126–27
voles, 7–8, 36, 39–40, 44–45, 76, 125

Wasatch Mountains, 144
Wengeler, Bill, 32, 41, 233
Weselmann, Bob, 207
West, Elena, 126
Whitbeck, Anne, 22
willows, 126
Wilmers, Chris, 133
winter, 5–6, plate 5
wolf 2. *See under* Leopold wolf pack
wolf 7. *See under* Leopold wolf pack
wolf 8. *See under* Rose Creek wolf pack
wolf 9. *See under* Rose Creek wolf pack
wolf 18. *See under* Rose Creek wolf pack
wolf 21. *See under* Druid Peak wolf pack
wolf 38. *See under* Druid Peak wolf pack
wolf 40. *See under* Druid Peak wolf pack
wolf 42. *See under* Druid Peak wolf pack
wolf 103. *See under* Agate Creek wolf pack
wolf 105. *See under* Buffalo Fork wolf pack
wolf 106. *See under* Geode Creek wolf pack
wolf 113. *See under* Agate Creek wolf pack
wolf 194. *See under* Specimen Ridge wolf pack
wolf 214. *See under* Nez Perce wolf pack
wolf 215. *See under* Nez Perce wolf pack
wolf 217. *See under* Slough Creek wolf pack
wolf 218. *See under* Buffalo Fork wolf pack
wolf 224. *See under* Druid Peak wolf pack
wolf 251. *See under* Agate Creek wolf pack
wolf 252. *See under* Nez Perce wolf pack
wolf 253. *See under* Druid Peak wolf pack
wolf 254. *See under* Druid Peak wolf pack
wolf 255. *See under* Druid Peak wolf pack

wolf 261. *See under* Slough Creek wolf pack
wolf 286. *See under* Druid Peak wolf pack
wolf 294. *See under* Geode Creek wolf pack
wolf 300. *See under* Geode Creek wolf pack
wolf 301. *See under* Leopold wolf pack
wolf 302. *See under* Leopold wolf pack
wolf 376. *See under* Druid Peak wolf pack
wolf 377. *See under* Slough Creek wolf pack
wolf 378. *See under* Slough Creek wolf pack
wolf 379. *See under* Slough Creek wolf pack
wolf 472. *See under* Agate Creek wolf pack
wolf 480. *See under* Leopold wolf pack
wolf 911, 80
Wolf Restoration Project: assisting people to see wolves, 33–34, 56–57, 67, 127–30; early-winter and late-winter studies, 12–13, 190; establishment of, 56; road crossings and road safety lessons, 32–33, 34–35, 37, 45, 100, 102; wolf population estimate, 193–94
wolves: assisting people to see, 56–57, 67, 127–30; bedding in snow, 7; cooperative behavior, 29, 45–46, 49–50, 75–76; deaths from fighting between rival packs, 141–42; hunting advantages against elk, 48, 163–64; in Japan, 129; loyalty among, 146–47; maintaining distance from, 130; naming of uncollared wolves, 131; Native Americans and, 50, 67, 127–28, 129; pack formation, 99; radio collars, 8–9, 63, 166; ravens and, 130–33; thinking by, 87–88; travel through snow, 49, 161–62; vegetation restoration and, 126–27; Yellowstone population, 193–94. *See also* Druid Peak wolf pack; food and feeding; hunting; play; pups; *other wolf packs*
Wolves on the Hunt (Mech, Smith, and MacNulty), 72–73
Wyman, Travis, 77

Yellowstone, 5–6, 193–94. *See also* Wolf Restoration Project
Yellowstone Delta wolf pack, territory, 2, *52*, *92*, *150*, *196*
Yellowstone Forever, 236

Zieber, Tom, 22, 23, 27, 32, 34–35, 36, 45, 89